Lecture Notes in Control and Information Sciences

Edited by A.V. Balakrishnan and M. Thoma

17

O. I. Franksen · P. Falster
F. J. Evans

Qualitative Aspects of Large Scale Systems

Developing Design Rules Using APL

Springer-Verlag
Berlin Heidelberg New York 1979

Authors

O. I. Franksen and P. Falster
Electric Power Engineering Department
The Technical University of Denmark

F. J. Evans
Department of Electrical & Electronic Engineering
Queen Mary College, University of London

ISBN 3-540-09609-4 Springer-Verlag Berlin Heidelberg New York
ISBN 0-387-09609-4 Springer-Verlag New York Heidelberg Berlin

Printing and binding: Beltz Offsetdruck, Hemsbach/Bergstr.
2060/3020-543210

FOREWORD

This monograph has been given a rather broad general title, although a glance at the table of contents quickly shows that it is constructed around the central concepts of controllability and observability. If any excuse is needed for this, it is that we felt the work does represent an investigation in the true inductive spirit from the particular to the general. We also feel that the use of an interactive APL facility has allowed the work to proceed in the classical empirical manner, which demonstrates the role of a "computer laboratory" in modern systems analysis.

As Karl Popper has pointed out, theories can never be verified but only falsified, by experiment, but it is experiment, fertilised by intuitive and imaginative thoughts, that generate the theories.

We have also been greatly influenced by several other convictions. Firstly, that the application of a group theoretic approach to systems analysis should bring benefit, and we have attempted to show that the adoption of a type of Erlanger Program which has proved so successful elsewhere, is also of value here. Indeed, a fundamental advantage of such an approach is that it separates qualitative properties from quantitative and yet combines them into a consistent logical framework permeated by abstract symmetries.

Secondly, we feel there is a need to supplement the algebraic symbolism and associated reaching after complete mathematical rigour, by the exposure of more qualitative or structural properties. This can be a source of further theoretical development in itself. Thirdly, it is our opinion that the type of approach described here is far more fruitful in the production of prescriptive design rules than the more analytic algebraic approach so far proposed.

So, in conclusion, we feel that these lecture notes will interest those who like to learn from numerical illustrations and are interested also in the scaffolding, which so often is carefully demolished in many current publications.

Lyngby and London, February 1979

O.I. Franksen, P. Falster & F.J. Evans

ACKNOWLEDGEMENT

The authors wish to express their gratitude
to The British Council for financial support
enabling them to overcome the problem of geo-
graphical distance, and their indebtedness to
Kommunedata I/S, the computer organization of
the Danish Municipalities, for making available
the excellent service of its APLSV system.

APL GLOSSARY

For convenience this glossary list the subset of build-in, socalled *primitive*, APL operations applied in this monograph. Many of these APL operations are introduced, illustrated, and explained in the course of the text. However, since the aim is to use APL as a tool, no attempt as such is made to actually introduce APL as a programming language or a mathematical notation. For this purpose the reader should consult other literature like the references given in the text or the programming manuals of the computer manufacturers.

For each of the APL operations of this glossary is given in tabular form: its name; APL notation; conventional mathematical notation (if it exists); a brief explanation; and a reference to an example in the text. In this connection it should be noted that the order of execution of an APL expression is *from right to left with no priority* among the operations. That is, parentheses are used in the conventional manner to introduce any desired priority order of execution. From an overall point of view the subset of APL operations listed here, may be subdivided into operations on the *shapes* of the arrays and operations on the *elements* of the arrays respectively. This classification is maintained in the glossary organizing in order the operations into two tables called Tables A1 and A2.

NAME	APL SIGN	MATH. SIGN	EXPLANATION	EXAMPLE
ORIGIN	$\Box IO \leftarrow 1$		Indexing beginning with 1 (or 0) as specified	
SIZE	ρA		Dimension or size of array A. $\rho\rho A$ gives the *rank* of A.	(I.11) (I.27)
RESHAPE	$V\rho A$		Reshapes right argument, array A, to the size of the left argument, vector (or scalar) V.	*ASSIGN* page 62
INDEX	$V[I]$ $M[I;J]$ $M[I;]$	V_i $M_{i,j}$ $M_{i.}$	Indexing of a vector V, a matrix M, and a row in matrix M.	(I.46)
COORDINATE	$...[I]A$		Following a primitive APL-operation (or function) this expression indicates that the operation is applied along the I'th coordinate axis of array A.	(I.68) (I.14a)
MONADIC TRANSPOSE	$\lozenge A$	A^t	Generalized transpose reversing the order of coordinate axes of array A. Note, that APL cannot distinguish between row and column vectors.	(I.18)

Table A1. Structural Operations Based on Index-Sets

NAME	APL SIGN	MATH. SIGN	EXPLANATION	EXAMPLE
DYADIC TRANSPOSE	$I\lozenge A$		A permutational transposition of the coordinate axes of array A specified by index vector I. Equating two indices in vector I produces a diagonal hyperplane of array A.	(I.14b) (I.26)
CATENATE	$A,[I]B$		Joining arrays A and B along an *existing* coordinate axis I, maintaining the maximal given rank.	(I.14a)
LAMINATE	$A,[K]B$		Joining arrays A and B along a *new* coordinate axis K (non-integer), increasing the rank by 1.	(I.14a)
TAKE	$V\uparrow A$		V is an integer or a vector of integers. If $V[I]$ is positive (negative), take the first (the last) $V[I]$ components of array A along its I'th coordinate axis.	*PLUSΔMPLY* page 9
COMPRESS	$V/[I]A$		V is a Boolean vector of the size of dimension I of array A. The components along dimension I of A corresponding to unit elements in V, are selected.	*PLUSΔMPLY* page 9

Table A1 continued

NAME	APL SIGN	MATH. SIGN	EXPLANATION	EXAMPLE
ASSIGNMENT	$R \leftarrow A$	R=A	Variable R defined equal to A.	(I.20)
OUTPUT	$\square \leftarrow A$		Assign the value of A to $\square$, i.e. print the value of A.	(I.24)
	A		Abbreviated version of the latter above.	(I.13)
MULTIPLE ASSIGNMENTS	$\square \leftarrow R \leftarrow A$		Printing variable R defined equal to A.	(I.24)
ORDER OF EXECUTION			Composite expressions are evaluated from right to left with no priority order among the operations	
PARENTHESES	(....)	(....)	Parentheses are used conventionally to introduce a priority order in the evaluation of computer expressions	(I.11)
ADDITION	$A \oplus B$	$A \oplus B$	Scalar "addition" generalized to the element-by-element operation $\oplus$ of conforming arrays A and B	(I.49)
REDUCTION	$\oplus/[I]A$	$\sum\limits_{i=1}^{N} A_{..i..}$	Scalar "summation" generalized to $\oplus$ reduction along the I'th coordinate axis of array A	(I.26) (I.68)
SCAN	$\oplus\backslash[I]A$		Scalar "accumulation" generalized to a $\oplus$ scan along the I'th coordinate axis of array A. Applied to a vector the result is a vector of the same size with the K'th element equal to the $\oplus$ reduction over the first K elements.	*ASSIGN* page 62

Table A2. Constituent Operations Based on Scalar Elements

NAME	APL SIGN	MATH. SIGN	EXPLANATION	EXAMPLE
OUTER PRODUCT	$A\circ.\otimes B$	$C_{mnpq} = A_{mn}B_{pq}$	Tensor outer product, evaluated as the "product" of each element in array A with each element in array B.	*PLUSΔMPLY* page 9
INNER PRODUCT	$A\otimes.\otimes B$	AB or $A\times B$	Tensor inner product generalizing the conventional matrix product to conforming arrays A and B.	(I.26) (I.67)
INVERSE	⊟M	M^{-1}	The inverse of matrix M or, if it does not exist, a *DOMAIN ERROR* message	(I.27)
TENSOR CONTRACTION	$+/[K]I\Diamond A$	A_k^k	Equating two indices K in the index-set I followed by a plus-reduction (summation) along the equated index K.	(I.20) (I.26)
INDEX GENERATOR	ιN		Applied to a non-negative integer N a vector of the first N integers is produced counting from origin □IO.	*PLUSΔMPLY* page 9
GRADE UP (DOWN)	$\blacktriangle V$ $(\blacktriangledown V)$		The index permutation vector that would sort the elements of vector V in ascending (descending) order.	(III.11)

Table A2 continued

TABLE OF CONTENTS

Chapter 1.   BACKGROUND AND SCOPE

Engineering design is the building upward of systems from available knowledge about the separate parts. The impact of the systems approach upon this process is a *formal awareness of the interactions* between the system parts. Basically, a system is conceived as an assembly of distinct parts, the socalled elements, the characteristics of which can be described without reference to the rest of the system. That is, the overall function of a system is determined, partly by the particular characteristics of each element, and partly by the manner in which the various elements are strung together.

The increase in size and complexity combined with the need for taking non-linearity into account, call for new concepts and procedures in the design process of automatic control systems. Mere knowledge of the system parts taken separately is not sufficient. On the contrary we can expect that, often, the interconnections and interactions between the system elements will be more important than the separate elements themselves. That is, the internal structure of a complex system is depicted, not only by its multitude of elements, but by its pattern of interconnections and interactions between elements. Therefore, an important aim in automatic control research must be to emphasize the *development of new, explicit, and highly organized design techniques for dealing with system structures* (Franksen, 1972; van Dixhoorn & Evans, 1974).

To meet this need for a maximum rationalisation of behaviour in the design process research must be undertaken to develop *qualitative or structural design rules* which, in the form of computer-aided methods, may be applied in the design of high-order and non-linear automatic control systems *without actually attempting explicit solutions*.

In this monograph we are concerned with a set of design concepts that represents, in a qualitative manner, the properties of controllability and observability of time-invariant linear systems. In classical mechanics the term *potential energy* designates the latent capacity of a system for doing mechanical work. Here, similarly, the design concepts under discussion will be called *potential controllability* and *potential observability* since they describe those latent qualities that make a system structurally amenable to control or observation. Their main virtue as design tools lies in their organized handling of large numbers of state-variables defining systems of high order or high dimensionality. However, besides curbing the problem of high dimensionality they are expected to apply also to non-linear systems by virtue of their simple, yet fundamental topological nature. Though, at the present stage of development this possibility must be considered merely a working hypothesis.

Basically, the concept of potential controllability and potential observability may be derived from a purely algebraic as well as from a purely topological or geometrical point of view as the necessary conditions that a time-independent linear system will be controllable respectively observable. The first to consider this problem seems to be Lin who, in the spirit of the wellknown planar graph theorem of Kuratowski, formulated two forbidden subgraphs making a single-input linear system, as he termed it, not *structurally controllable* (Lin, 1974). Turning from Lin's geometrical or graph-theoretic viewpoint to a purely algebraic approach Shields and Pearson extended his results to multi-input linear systems (Shields & Pearson, 1976). In particular, they identified one of Lin's two forbidden subgraphs with the algebraic determination of the socalled *term rank* of a matrix (i.e., the maximal rank a matrix may attain by virtue of its structural pattern of non-zero elements). Though, unaware of the accepted mathematical terminology, they somewhat confusingly designated this concept the *generic rank* of a matrix. Also, they developed an algorithm based on permutation operations to ascertain this rank. Their algebraic approach, subsequently, was considerably simplified by Glover and Silverman who realized that the other of Lin's two forbidden subgraphs represented purely logic properties that could be determined recursively by Boolean matrix operations only (Glover & Silverman, 1976). Recently, extending the approach to the realm of observability,

Davison switched back to Lin's graph-theoretic point of view and gave these Boolean operations an interpretation in terms of the *connectability* properties of a *linear graph* (Davison, 1977). In fact, though, what he considered was not the connectability of a linear graph, but the socalled *reachability* of a different kind of graph known as *directed graphs* or *digraphs* for short (Franksen, 1975).

Independently of the abovementioned contributions and simultaneously with those of Shields & Pearson and Glover & Silverman the present authors developed the qualitative design concepts of potential controllability and potential observability from the combined viewpoint of systems engineering and control theory (Franksen, et al., 1976). Theoretically, this work was founded on a representation of measurements and systems structures by Cartesian and Boolean tensors respectively by digraphs (Franksen, 1968; 1975; and 1976). Indeed, instead of considering the tensorial and graph-theoretical approaches alternative to each other, we consider them complementary. That is, the purpose of a tensorial formulation is *system aggregation*, whereas the aim of a graph-theoretical description is *system decomposition*. Evidently, both of these aspects must be taken into account in the practical design of a system. However, also from a theoretical point of view it is important not to neglect the one approach in favour of the other since they tend to bring out different system properties. To illustrate, the tensorial formulation will emphasize the inherent symmetries of the problem, while the graph-theoretical description will identify the distinct structural characteristics underlying a specific set of quantitative properties. Thus, in the precursor of the present book the former approach led the authors to the discovery of a neglected duality, designated input observability, while the latter approach gave rise to a formulation of the tensorial properties of potential controllability and potential observability in terms of the socalled reachability matrix of a digraph (Franksen, et al., 1976).

By comparison with Lin's contribution the tensorial approach, adopted by the authors, represented a generalization of only one of his criteria. That is, only the structural properties depicted graph-theoretically by the reachability matrix will appear as genuine tensor properties. In contradistinction the structural properties generalized by Shields and Pearson from Lin's other criterion, are additional non-tensorial properties related solely to the question of matrix inversion. Indeed, the determinantal problem of matrix rank, made qualitative by the term rank determination, is excluded from any purely tensorial consideration. It follows that potential controllability or potential observability of a time-independent linear system must be established by the simultaneous satisfaction of two independent conditions which for convenience we shall call respectively *the reachability test* and *the term rank test*.

The very fact that a tensorial approach will exclude any consideration of the non-tensorial properties, explains why two independent conditions must be satisfied. Yet, Lin's approach indicates that both criteria may be dealt with in a uniform manner from a graph-theoretic point of view. In this respect, perhaps, it is obvious that the interpretation of the contracted tensors in terms of the reachability matrix will provide the necessary digraph interpretation of the tensorial properties. However, it is not so evident that we must invoke the lesser known concept of alternating paths to obtain a digraph representation of the non-tensorial property of term rank.

From this description it will appear that the concepts of potential controllability and potential observability, may be discussed from either a purely tensorial or a purely graph-theoretical point of view. Though, we may also combine these two viewpoints contrasting them in the sense of representations of system aggregation and system decomposition. All together, therefore, three possibilities of presentation open up, each complementing the other two. This accounts for the organization of this book into three parts dealing respectively with the tensorial, the graph-theoretical, and the combined viewpoints. Briefly, the contents of these three parts may be stated as follows.

In part I the conventional matrix criteria for quantitative investigation of the controllability and the observability of linear, time-invariant systems are recast into dual invariant forms stated in terms of the component arrays of Cartesian tensors. In the spirit of Klein's Erlanger Program the structural contents of these criteria, maintaining the invariant form, are then given a topological representation based on the component arrays of corresponding, more general Boolean tensors. Thus, by purely algebraic considerations, the tensorial properties of the qualitative design concepts of potential controllability and potential observability are established.

In part II the Boolean tensor criteria are given an alternative formulation in terms of invariant properties of digraphs. Interpretation of the state equation coefficient matrix as a digraph, depicting system states as vertices, furnishes a simple explanation of the qualitative design concepts in terms of fundamental reachability properties between digraph vertices. Based on this interpretation these design concepts may be derived from the system digraph either directly by visual inspection or by a simple set of corresponding Boolean matrix operations. The importance of the latter matrix operations is twofold. They are computationally simple, yet theoretically fundamental. This is due to the fact that they establish a Boolean isomorphism to the wellknown quantitative criterion for state controllability and observability originally proposed by Gilbert.

By this isomorphism Gilbert's assumption of distinct eigenvalues is turned into a Boolean assumption on the socalled term rank of the coefficient matrices. Essentially, the term rank of a matrix is its maximum number of independent entries. In a Boolean representation the pattern of these entries forms a maximal permutation matrix that may be derived from a graph-theoretical point of view by consideration of the socalled alternating paths of the corresponding bipartite digraph. Establishment of a defining set of alternating paths may take place either by visual inspection or by an exhaustive algorithmic search which, by the way, may be conceived also as a solution procedure to the wellknown assignment problem of combinatorial mathematics. It is noteworthy that term rank, in contradistinction to digraph reachability, is not a Boolean tensor property since alternating paths are invariant only under the alternating group or the group of even permutations. Of course, this makes the analogy even more striking between Gilbert's quantitative approach to state controllability and observability and the qualitative digraph approach adopted here to potential controllability and observability.

In part III the two viewpoints, the tensorial and the graph-theoretical, are considered side by side as complementary in aim to each other. Thus, just as system decomposition springs from the digraph formulation, so system aggregation ties in with the tensor formulation. Beginning with the former, the reachability matrix is submitted to a Boolean counterpart of the wellknown partitioning operation of a quantitative matrix into its symmetrical and antisymmetrical component parts. Submitting the antisymmetrical component of the reachability matrix to a conventional level-coding opens to a novel, and more fundamental, explanation of the concept of quasi-levels. Though, of course, to determine the latter a corresponding structural decomposition of the digraph into its strong cyclic components and its condensed acyclic digraph must be performed. Based on the outcome of this operation the systems equation may be blocked into a hierarchy of independent subsets each of which may be dealt with separately with respect to determination of the various quantitative properties of the system. Obviously, the structure of some systems is such that they are not decomposable as will be identified submitting such systems to the abovementioned procedure. This, however, does not diminish the importance of the approach presented since in practice we usually find that higher order systems are decomposable.

From the tensorial point of view a universal tensor of rank 3 may be formulated on the basis of an aggregation of the state and output equations. The characteristic feature of this tensor, is that it combines all the quantitative rank tests of controllability and observability into a single entity.

In particular, it is noteworthy that this tensor will encompass also the additional concept of output controllability in such a way that, by symmetry considerations, a new dual concept, called input observability, may be formulated. Turning to a digraph interpretation of the latter set of dual concepts their counterpart structural or qualitative design concepts are established. To specify these concepts of potential output controllability and potential input observability a new pair of defining reachability and term rank tests are formulated. Thus, by the tensor aggregation all the inherent symmetries of the system are brought out in its mathematical formulation.

From this brief review of the contents of this monograph it will be evident that we are concerned exclusively with a set of structural or qualitative design concepts called potential controllability and potential observability. However, by our choice of notation, and our way of presentation, we desire to place our discussion in the framework of a more general approach which we believe of importance in the search for qualitative design rules beyond the scope of this book. That is, we have based our formulation entirely on the interactive computer language APL, which is an acronym for "A Programming Language" (Iverson, 1962; and Pakin, 1968).

This choice has been made for two reasons. First, no other standardized mathematical notation exists for the explicit and unique representation of generalized algebraic operations on multi-dimensional arrays. Secondly, the interactive implementation of APL allows for a direct computer execution of the given mathematical formulation. Hence, adopting APL enables us to use the computer as a laboratory for performing numerical experiments (Franksen, 1977). Accordingly, a secondary object of the presentation is to demonstrate the idea of an experimental development by computer of theoretical models. To achieve this we have, on the one hand, collected all the pertinent results in tables which, formulated in APL, will enable the reader directly to make the approach operational for computer execution. On the other hand, we have emphasized the operational aspects of the proposed approach by using terminal outputs, executed interactively on an APL-terminal, as numerical illustrations. The reader unfamiliar with APL may be referred to the APL Glossary given at the outset of this monograph. This glossary explains some of the more important APL operations to be used in the following. Also, it may serve as a ready reference during the reading of the following text. Finally, lest we should be misunderstood we would like to emphasize that we consider APL not a substitute, but a useful complement to the conventional mathematical formulation.

PART I: THE TENSORIAL APPROACH

INTRODUCTION

A natural starting point for the tensorial approach to state controllability and observability will be the conventional matrix criteria for quantitative investigation of these dual properties of time-invariant linear systems. Hence, to fix ideas, we shall briefly review these matrix criteria. The importance of this review, though, goes beyond the usual definition of the notation of the equations and the dimensions of their constituent parts. In fact, we shall deviate from the conventional formulation by taking out the power series of the state equation coefficient matrix as a separate entity. Seemingly, the isolation of this entity is an unnecessary complication of no importance. Yet, it is this insignificant step that paves the way for a tensorial formulation of the entire problem.

The power series of the state equation coefficient matrix give rise to a Cartesian tensor of rank 3 specifying the system properties. Performing the inner product of the component array of this tensor with the appropriate coefficient matrices of respectively the state and output equations, produces two tensors of rank 3. The properties of state controllability and observability are governed each by one of these tensors called respectively the state controllability tensor K and the observability tensor M. Tensor contractions on the inner product of the transpose of M and M, yield two matrix components submitting to the conventional rank tests. An important theoretical consequence of this formulation of the rank tests, is the establishment of a distinct symmetry which, conceived as an invariant system property, opens up for the introduction of a new concept dual to that of output controllability. This concept, to be discussed in part III, will be called input observability.

The Cartesian tensor formulation, maintaining the invariant form, is then given a topological representation based on the component arrays of corresponding, more general Boolean tensors. Thus, two Boolean tensors of rank 3 are established as the structural counterparts of the state controllability tensor and the observability tensor. Submitting the component arrays of these tensors to the existential quantifier of Boolean algebra, produces two matrix components governing the structural properties of the system. Hence, by purely algebraic considerations, the tensorial properties of the design concepts of potential controllability and potential observability are established.

By its very nature the tensorial approach, to be undertaken in the present part I, will completely neglect the term rank test discussed previously. In fact, to simplify the discussion we shall find it convenient to *assume that the term rank condition holds everywhere*. Of course, this assumption does not relate to the formulation of the tensorial properties since they are independent of the non-tensorial properties defined by the term rank test. Rather, its only purpose is to permit us to carry out numerical comparisons with the conventional quantitative rank tests for state controllability.

Chapter 2. THE CARTESIAN TENSOR FORMULATION

In this chapter, to fix ideas, we shall briefly review the conventional matrix criteria for quantitative investigation of the dual properties of state controllability and observability of linear, time-invariant systems. These criteria are then recast into dual invariant forms stated in terms of component arrays of Cartesian tensors. The identification of the power series of the state equation coefficient matrix as a Cartesian tensor Q of rank 3, is the conceptual foundation on which this approach must be based. In fact it is by means of this tensor that the two conventional criteria are redefined as properties of two derived Cartesian tensors, the state controllability tensor K and the observability tensor M. Thus, this reformulation permits a subdivision of the design procedure into two steps: a topological or structural phase and an algebraic or quantitative phase. Here, as in the remainder of this book, the presentation will be primarily operational, that is, formulated in APL (see the APL Glossary) and illustrated numerically by interactive computer terminal outputs. The numerical example, applied throughout, is a fourth-order system with the constant coefficients taken at simple integer values to make it easy to follow the computations.

2.1 The Basic Criteria

Assume that a linear, time-invariant system is given a mathematical representation in terms of two matrix equations designated respectively the state equation and the output equation.

- *State Equation:* $\qquad\qquad \dot{X} = A{\times}X + B{\times}U \qquad\qquad$ (1)
- *Output Equation:* $\qquad\quad\; Y = C{\times}X + D{\times}U \qquad\qquad$ (2)

In these equations X is a vector of X state variables, $\dot{X}{=}dX/dt$ its time-derivative, U a vector of U input or control variables, Y a vector of Y output variables, and A, B, C, and D constant coefficient matrices which conform in dimensions with these vectors. Considering d/dt as a scalar operator on vector X we may depict the state and output equations graphically as indicated in Fig. 1 if we wish to emphasize the dimensions of the arrays involved.

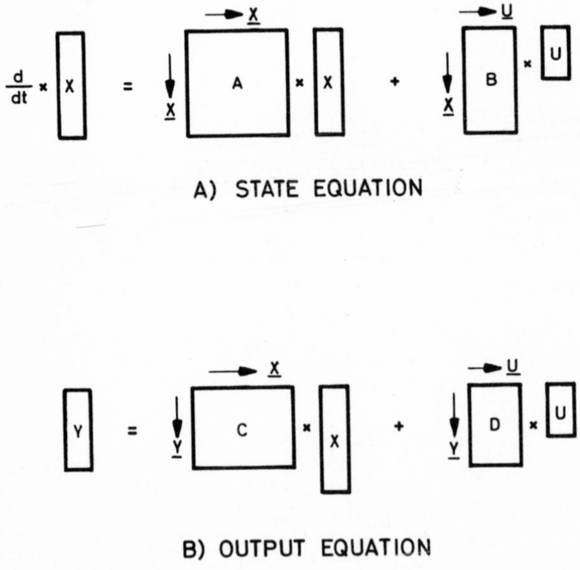

A) STATE EQUATION

B) OUTPUT EQUATION

Fig. 1.   State and Output Equations

Now, for a system thus defined, the dual concepts of controllability and observability have been introduced as computational design aids that may be used to verify whether or not a proposed system configuration will possess certain desirable properties (Kalman, 1963; Gilbert, 1963). Essentially, the question of the state controllability of a system reduces to the fact whether or not it is possible, by means of the independent or unbounded control variables U, to drive in finite time any one of the state-variables X from an arbitrary initial state to a desired state. If this is possible for every state-variable the system is said to be *completely state controllable*. The concept of observability is dual to that of controllability in that it depends upon the output variables Y instead of the input variables U. Basically, the question of observability amounts to the fact whether or not measurements of the output Y contain sufficient information to identify, in finite time, any state X. In other words, a system is *completely observable* if every state-variable of the system affects some of the outputs.

The necessary and sufficient matrix criteria for investigation of whether a system is completely state controllable or completely observable, are based solely on the properties of the coefficient matrices A, B, and C of the state and output equations. Thus, the coefficient matrix D, relating the control variables U with the output variables Y, does not influence this investigation. But, as we shall see later, matrix D enters the stage in a derived criterion which is used to determine an additional property of the system called output controllability. A system is *completely output controllable* if it is possible, by means of the independent control variables U, to drive in finite time any arbitrary output Y to any desired final set of values. For the present, however, we shall be concerned solely with the criteria of complete state controllability and observability.

The criterion for complete state controllability derives from the state equation (1). Thus, it is based only on the coefficient matrices A and B. Recalling that $\underline{X}$ is the dimension of the vector X of state variables, we shall find it useful to introduce a scalar:

$$P = \underline{X} - 1 \tag{3}$$

designating the highest power of the following matrix power series of A:

$$Q = A^0, A^1, A^2, \ldots, A^P \tag{4}$$

where, as usual, A^0 is a unit matrix. Now, postmultiplying each of these terms by matrix B, we may catenate these $\underline{X}$ terms into a new matrix K of dimensions $(\underline{X}, \underline{X} \times \underline{U})$:

$$K = \{B, A \times B, A^2 \times B, \ldots, A^P \times B\} \tag{5}$$

It is wellknown that a necessary and sufficient condition that the system is completely state controllable, is that the rank of this matrix K equals $\underline{X}$. Hence, by testing the rank of K we test the state controllability of the proposed system configuration. It should be observed in this connection that, as it stands, the criterion gives no hints on how to actually design the system.

The criterion for complete observability derives from the state and output equations together (1 & 2) in that it is based solely on the coefficient matrices A and C. Again we develop the power series Q in (4) of A. Introducing C instead of B the dual operation of that applied for state controllability, is to premultiply each of the terms in Q by C and catenate these $\underline{X}$ terms below each other into a new matrix M of dimensions $(\underline{X} \times \underline{Y}, \underline{X})$:

$$M = \left\{ \begin{array}{c} C \\ C \times A \\ C \times A^2 \\ \cdot \\ \cdot \\ C \times A^P \end{array} \right\} \qquad (6)$$

As before it may be shown that a necessary and sufficient condition that the system is completely observable, is that the rank of matrix M equals X. Again we realize that the criterion, in this formulation, offers no clue on how to redesign a proposed system configuration that fails to meet the requirement of complete observability.

Computationally both criteria hinge on the determination of the rank of a rectangular matrix K or M. One method of deciding if, say, X=rank K is to see if the determinant of K postmultiplied by its transpose is non-zero:

$$\det (K \times K^t) \neq 0 \qquad (7)$$

since rank K = rank $(K \times K^t)$. Of course, in the dual case of matrix M the determinant to be considered, is $\det(M^t \times M)$. In practice other and faster computer algorithms exist for rank determination, but for our purpose this approach like (7) is not without merit.

2.2 A Tensorial Power Series Representation

A fine, yet important distinction between the above presentation and that of the current literature in the formulation of the criteria for complete state controllability and observability, is the introduction in (4) of the power series Q as an intermediate step. Clearly, Q must describe some intrinsic, indeed *invariant* properties of the system configuration. Properties, that characterize the structural relationships between the possible system states X independent and distinct from any additional relationships to input variables U or output variables Y. The very idea that Q represents the invariant physical nature of the system, suggests that we should conceive Q as a tensor. Now, the dimension of matrix A is (X,X) where X is the number of state-variables X. Also, the number of powers of A in Q is X, namely, the powers $0,1,2,\ldots,P$ with P defined by (3) to be one less than X. However, in principle it is possible to introduce any arbitrary number P of powers of A. In other words, the number of state-variables X is limited by the physical structure of the system to a total of X, but no such a priori limitation is imposed upon the number P of powers of A. Hence, the dimension of Q is that of A with an additional coordinate axis for the powers of A. That is, Q is measured in a three-dimensional space spanned by, as the most natural choice, three rectangular or Cartesian axes. A little reflection on the possible transformations of this space, introducing other state variables or permuting the powers of A, will show that Q must be invariant under such transformations. Accordingly, *Q is a Cartesian tensor of rank 3*. The three dimensions of Q are X,X and P where, in relation to the criteria of controllability and observability, we have P=X. Thus, in any given coordinate system the Cartesian tensor Q is represented by a 3-dimensional or box-like component array of coordinates. Of course, these coordinates will change under any permissable transformation of the coordinate system.

From a computational point of view all we can determine of Q is its 3-dimensional component array of coordinates in any given reference system. Essentially, this reference system is fixed by the given coefficient matrix A of the state equation. In APL the individual terms of the power series of A, say, the

J'th power of A, denoted AJ, is determined:

$$AJ \leftarrow A+.\times A+.\times A+.\times....+.\times A \qquad (8)$$

where the APL inner product "$+.\times$" specifies the conventional matrix product of the J terms in (8). A recursive function, valid in either APL origin (see APL Glossary), for calculation of the J'th power of A is:

```
    ∇ PLUSΔMPLY [□] ∇
  ∇ R←M PLUSΔMPLY I
[1]  ⍝
[2]  ⍝      THE NON-NEGATIVE AND INTEGRAL POWER
[3]  ⍝      "I" OF A SQUARE MATRIX "M".
[4]  ⍝
[5]     →(I=0)/END
[6]     R←M+.×M PLUSΔMPLY I-1
[7]     →0
[8]  END:R←(⍳1↑⍴M)∘.=⍳1↑⍴M
  ∇
```

Hence, the J'th power of A is found in general as:

$$AJ \leftarrow A \; PLUS\Delta MPLY \; J \qquad (9)$$

with the 0'th power of A, denoted $A0$, computed as a unit matrix by statement [8] in this APL function.

In APL we use the primitive function ρ to designate the dimension or rank of any array. For example, the dimensions of matrix A are:

$$(\underline{X},\underline{X}) = \rho A \qquad (10)$$

To preserve the symmetry between the dual criteria for state controllability and observability we shall find it useful to define the dimensions of the component array of the Cartesian tensor Q by the expression:

$$(\underline{X},\underline{P},\underline{X}) = \rho Q \qquad (11)$$

This choice of order of the coordinate axes being guided by the fact that it should be immediately possible to carry out a postmultiplication by B and a premultiplication by C.

With the different powers of A calculated by (9), assuming 1-origin, we may establish the component array of Q as follows:

$$Q \leftarrow A0,[2]A1,[2]....,[1.5]AP \qquad (12)$$

As illustrated in Fig. 2 this component array of Q preserves the $\overline{P}$ powers of A as an invariant entity at the same time as it opens up for an individual, yet simultaneous investigation of the different powers.

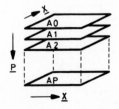

Fig. 2. Component Array of the Cartesian Tensor Q

As a numerical example consider the following coefficient matrix A of a fourth order system:

$$
A
$$
$$
\begin{array}{cccc}
0 & -1 & 0 & 0 \\
0 & -2 & 0 & 0 \\
-0 & 0 & 0 & 1 \\
-5 & 0 & 5 & 0
\end{array}
\tag{13}
$$

Since X=4 we have by (3) that the four powers of interest are the powers 0,1, 2, and 3. These are calculated by (9) as follows:

$$\square \leftarrow A0 \leftarrow A \; \underline{PLUS \triangle MPLY} \; 0$$

$$
\begin{array}{cccc}
1 & 0 & 0 & 0 \\
0 & 1 & 0 & 0 \\
0 & 0 & 1 & 0 \\
0 & 0 & 0 & 1
\end{array}
$$

$$\square \leftarrow A1 \leftarrow A$$

$$
\begin{array}{cccc}
0 & -1 & 0 & 0 \\
0 & -2 & 0 & 0 \\
-0 & 0 & 0 & 1 \\
-5 & 0 & 5 & 0
\end{array}
$$

$$\square \leftarrow A2 \leftarrow A \; \underline{PLUS \triangle MPLY} \; 2$$

$$
\begin{array}{cccc}
0 & -2 & 0 & 0 \\
-0 & 4 & 0 & 0 \\
-5 & 0 & 5 & 0 \\
0 & -5 & 0 & 5
\end{array}
$$

$$\square \leftarrow A3 \leftarrow A \; \underline{PLUS \triangle MPLY} \; 3$$

$$
\begin{array}{cccc}
0 & -4 & 0 & 0 \\
0 & -8 & 0 & 0 \\
-0 & -5 & 0 & 5 \\
-25 & 10 & 25 & 0
\end{array}
$$

Applying (12) the component array of the Cartesian tensor Q may be found:

$$Q \leftarrow A0,[2]A1,[2]A2,[1.5]A3$$
$$\rho Q$$
$$4 \quad 4 \quad 4 \tag{14a}$$

Note, in this connection, that the dimensions of Q, ρQ, should be interpreted by (11).

Now, to obtain an output of Q in a form easy to verify we use the transposition 2 1 3$\lozenge Q$, as illustrated in Fig. 3, to turn the spatial arrangement of the component array before printing.

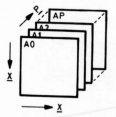

Fig. 3. The Transposed Component Array 2 1 3$\lozenge Q$

It is then easy to see that Q has been correctly specified:

$$
\begin{array}{cccc}
2 & 1 & 3 & \mathbb{Q}Q \\
\end{array}
$$

$$
\begin{array}{cccc}
1 & 0 & 0 & 0 \\
0 & 1 & 0 & 0 \\
0 & 0 & 1 & 0 \\
0 & 0 & 0 & 1 \\
\\
0 & 1 & 0 & 0 \\
0 & -2 & 0 & 0 \\
0 & 0 & 0 & 1 \\
-5 & 0 & 5 & 0 \\
\\
0 & -2 & 0 & 0 \\
0 & 4 & 0 & 0 \\
-5 & 0 & 5 & 0 \\
0 & -5 & 0 & 5 \\
\\
0 & -4 & 0 & 0 \\
0 & -8 & 0 & 0 \\
0 & -5 & 0 & 5 \\
-25 & 10 & 25 & 0 \\
\end{array}
$$

(14b)

2.3 The Controllability Criterion Reformulated

In physical terms controllability implies that it is possible, with a given set of input or control variables U, to completely change in accordance with our wishes the state of the system, i.e. the state variables X. The distinction in this statement between the system itself and its exogenous control is reflected mathematically by the separation in the state equation (1) of the coefficient matrices A and B. Thus, as the *intrinsic system properties* derive from matrix A in the form (12) of the Cartesian tensor Q, the built-in physical properties of *control* originate in matrix B considered as a two-dimensional subcomponent array of rank 2 of a "universal" Cartesian tensor to be introduced in part III. As indicated in Fig. 1 the dimensions of the component array are defined:

$$(\underline{X},\underline{U}) = \rho B \tag{15}$$

In the sense of a design concept controllability combines into a whole, the system properties with those of the control. To represent the invariant physical unity of the combination of the system with the control we invoke the inner product of Q and B resulting in a new Cartesian tensor K of rank 3, designated the *state controllability tensor*:

$$K \leftarrow Q + . \times B \tag{16}$$

The component array of this tensor, which appears instead of the conventional matrix formulation (5), is readily seen (by (11) and (15) as illustrated in Fig. 4) to have the dimensions:

$$(\underline{X},\underline{P},\underline{U}) = \rho K \tag{17}$$

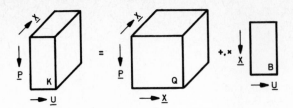

Fig. 4.   Component Arrays of the State Controllability Tensor

Based on the state controllability tensor K the rank determination test of the conventional state controllability criterion may be generalized as follows. First, adopting the idea of (7) we establish the 4-dimensional component array of the inner product of K and its generalized transpose $\lozenge K$ (i.e., K with the order of the dimensions in (17) reversed):

$$K+.\times \lozenge K \tag{18}$$

Evidently, the dimensions of this product are:

$$(\underline{X},\underline{P},\underline{P},\underline{X}) = \rho(K+.\times \lozenge K) \tag{19}$$

Secondly, we perform a tensor contraction on the repeated index P. That is, we equate the two indices P (taking out a 3-dimensional diagonal hyperplane of the product (18) as indicated in Fig. 5) and perform a summation or plus-reduction along the repeated index P. Redefining K to designate the outcome of this operation, the tensor contraction on (18) may be specified in 1-origin, reading as usual in APL from right to left, by the following expression:

$$K \leftarrow +/[3]1 \ 3 \ 3 \ 2\lozenge K+.\times \lozenge K \tag{20}$$

Obviously, the dimensions of K thus redefined will be:

$$(\underline{X},\underline{X}) = \rho K \tag{21}$$

In other words, we end up with a square matrix K. It is easy to show, as we shall illustrate later numerically, that this matrix K is identical with the matrix product in (7). Hence, the rank test for state controllability is now simply a matter of determining whether the determinant of K is non-zero.

In APL we may define some function to perform this test. However, for investigations of lower-order systems it is more convenient simply to attempt an inversion of K using the primitive operator "⊟". If indeed we obtain the inverted matrix ⊟K the system is state controllable. On the other hand, if the system is not state controllable we will invoke an APL system error message: "DOMAIN ERROR". For mnemonic reasons, since only the existence but not the actual value of the inverse of K is of interest, we shall perform this test using the rank operator ρ as follows:

$$\rho \boxminus K \tag{22}$$

Then, the system will return the dimensions of the inverse of K or an error message depending upon whether or not the system is completely state controllable.

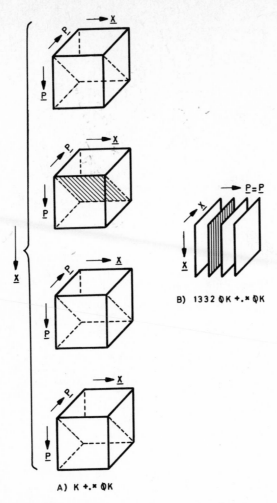

B) 1332 ⌽K +.× ⌽K

A) K +.× ⌽K

Fig. 5. Equating the Repeated Index $\underline{P}$

A special case of some interest occurs in the above approach whenever B is a vector of dimension X. In this situation, obviously, the dimensions of the component array of $\overline{K}$ defined by (16) are simply:

$$(\underline{X}, \underline{P}) = \rho K \qquad (23)$$

That is, the component array K is a square matrix since $\underline{P}=\underline{X}$. Therefore, whenever B is a vector we may skip (18) and (20) applying the $\overline{t}$est (22) directly to the outcome of (16).

Let us assume, to illustrate this approach numerically, that the fourth order system specified by the A-matrix of (13) is submitted to a control defined by a coefficient *matrix* B:

$$\square \leftarrow B \leftarrow B1$$

```
        1  0
        0  2
        0  0
        0  0
```
$$\rho B$$
```
        4  2
```
$$(24)$$

Introducing this control and the Cartesian tensor Q of (14) into (16), yields the state controllability tensor:

$$\square \leftarrow K \leftarrow Q + . \times B$$

$$\begin{array}{rr}
1 & 0 \\
0 & 2 \\
0 & {}^-4 \\
0 & 8
\end{array}$$

$$\begin{array}{rr}
0 & 2 \\
0 & {}^-4 \\
0 & 8 \\
0 & {}^-16
\end{array}$$

$$\begin{array}{rr}
0 & 0 \\
{}^-0 & 0 \\
{}^-5 & 0 \\
0 & {}^-10
\end{array}$$ \hfill (25)

$$\begin{array}{rr}
0 & 0 \\
{}^-5 & 0 \\
0 & {}^-10 \\
{}^-25 & 20
\end{array}$$

$$\begin{array}{ccc}
& \rho K & \\
4 & 4 & 2
\end{array}$$

the dimensions ρK of which should be interpreted by (17).

Applying (18) produces a 4-dimensional component array with the dimensions particularized by (19):

$$\begin{array}{cccc}
& \rho K + . \times & \text{\small Q} K & \\
4 & 4 & 4 & 4
\end{array}$$

Redefining K by (20) as the result of a tensor contraction on the repeated index $\underline{P}$ of the inner product of K by its generalized transpose, furnishes the symmetrical 2-dimensional array:

$$\square \leftarrow \underline{K} \leftarrow + / [3] \underline{1} \ 3 \ 3 \ 2 \text{\small Q} K + . \times \text{\small Q} K$$

$$\begin{array}{rrrr}
{}^-85 & {}^-168 & {}^-80 & 200 \\
{}^-168 & 340 & 160 & {}^-400 \\
{}^-80 & {}^-160 & {}^-125 & {}^-200 \\
200 & {}^-400 & {}^-200 & 1150
\end{array}$$ \hfill (26)

$$\begin{array}{cc}
& \rho K \\
4 & 4
\end{array}$$

with the dimensions specified by (21).

Finally, we submit the thus redefined square matrix K to the rank test of (22):

$$\begin{array}{c}
\rho \boxminus K \\
4 \quad 4
\end{array}$$ \hfill (27)

which, by returning the dimensions of the inverse, signifies that the system is completely state controllable.

A comparison of the revised tensorial approach with the conventional matrix formulation is readily performed. For example, based on (25), we may establish the K-matrix of (5) with dimensions $(\underline{X}, \underline{X} \times \underline{U})$ as follows:

$$\square \leftarrow K \leftarrow (4,4\times2)\rho Q+.\times B$$

$$
\begin{array}{cccccccc}
1 & 0 & 0 & 2 & 0 & {}^-4 & 0 & 8 \\
0 & 2 & 0 & {}^-4 & 0 & 8 & 0 & {}^-16 \\
0 & 0 & 0 & 0 & {}^-5 & 0 & 0 & {}^-10 \\
0 & 0 & {}^-5 & 0 & 0 & {}^-10 & {}^-25 & 20
\end{array}
\tag{28}
$$

$$\rho K$$
$$4 \quad 8$$

The inner product in (7) of this matrix by its transpose yields a square matrix of dimensions $(\underline{X},\underline{X})$:

$$K+.\times \mathbb{Q}K$$

$$
\begin{array}{cccc}
85 & {}^-168 & {}^-80 & 200 \\
{}^-168 & 340 & 160 & {}^-400 \\
{}^-80 & 160 & 125 & {}^-200 \\
200 & {}^-400 & {}^-200 & 1150
\end{array}
\tag{29}
$$

$$\rho K+.\times \mathbb{Q}K$$
$$4 \quad 4$$

but this result is identical with that of (26). Hence, the difference between the tensorial approach and the conventional matrix approach is primarily a matter of formulation rather than calculation. In fact, though, the tensorial approach is somewhat more costly in computing effort.

To round off this illustration of the testing for complete state controllability let us consider a situation whereby the system, represented by the Cartesian tensor Q of (14), is submitted to a control defined by a coefficient *vector* B:

$$\square \leftarrow B \leftarrow B2$$
$$0 \quad 0 \quad 3 \quad 0 \tag{30}$$
$$\rho B$$
$$4$$

Applying (16) creates a square matrix - with its dimensions defined by (23) - as the component array of the state controllability tensor:

$$\square \leftarrow K \leftarrow Q+.\times B$$

$$
\begin{array}{cccc}
0 & 0 & 0 & 0 \\
0 & 0 & 0 & 0 \\
3 & 0 & 15 & 0 \\
0 & 15 & 0 & 75
\end{array}
\tag{31}
$$

$$\rho K$$
$$4 \quad 4$$

Performing the test (22) directly on this matrix:

$$\rho \boxplus K$$
$$DOMAIN \ ERROR \tag{32}$$
$$\rho \boxplus K$$
$$\wedge$$

reveals that submitting the system to the control vector (30) does *not* make the system completely state controllable.

2.4 The Observability Criterion Reformulated

The physical meaning of question of observability is simply whether one can determine the initial values of all state variables X, given a sufficient number of measurements or output variables Y. Conceived as an abstract concept observability is the dual of controllability. Thus, considering the state and output equations (1 & 2), we see that in a similar manner as the coefficient matrix B specifies how the system is constrained with respect to the control or input variables U, so the coefficient matrix C specifies how the system is constrained with respect to the measurements or output variables Y. In other words, just as the built-in physical properties of control originate in matrix B considered as a sub-component array of some universal tensor of rank 2, so the inherent properties of the "output indicator" derive from matrix C apprehended as the sub-component array of rank 2 of the same tensor. The use of the word *output indicator* or, simply, *indicator*, is meant to imply only the structural relationship between available outputs and system states. From the illustration in Fig. 1 we see that the dimensions of the component array are defined:

$$(\underline{Y},\underline{X}) = \rho C \tag{33}$$

In the sense of a design concept observability integrates into a single unity the intrinsic system properties, represented by the Cartesian tensor Q (12), with those of the indication or measuring arrangement. To depict this integration as an invariant physical entity we invoke the inner product of C and Q resulting in a new Cartesian tensor M of rank 3, termed the *observability tensor:*

$$M \leftarrow C + . \times Q \tag{34}$$

The component array of this tensor, which replaces the conventional matrix formulation (6), is readily seen by (11) and (34) to have the dimensions:

$$(\underline{Y},\underline{P},\underline{X}) = \rho M \tag{35}$$

The rank determination test for complete observability, based on the observability tensor M, may be established by duality considerations from the corresponding test for state controllability. First, we introduce the dual of (18):

$$(\oslash M) + . \times M \tag{36}$$

Obviously, corresponding to (19) the dimensions of this inner product are:

$$(\underline{X},\underline{P},\underline{P},\underline{X}) = \rho (\oslash M) + . \times M \tag{37}$$

Secondly, corresponding to (20) we redefine M performing a tensor contraction on the repeated index $\underline{P}$:

$$M \leftarrow + / [3] 1 \ 3 \ 3 \ 2 \oslash (\oslash M) + . \times M \tag{38}$$

the outcome of which, corresponding to (21), is a square matrix of dimensions:

$$(\underline{X},\underline{X}) = \rho M \tag{39}$$

Finally, corresponding to (22) the rank determination test is performed by executing:

$$\rho \boxminus M \tag{40}$$

which, as before, will result in the dimension of the inverse of M or an error message depending upon whether or not the system is completely observable.

In the special case where C is a vector of dimension X, we find corresponding to (23) that the component array of M, defined by (34), is a square matrix of dimensions:

$$(\underline{P},\underline{X}) = \rho M \tag{41}$$

with $\underline{P}=\underline{X}$. Accordingly, whenever C is a vector we may neglect (36) and (38) applying the test (40) directly to the component array (34) of the observability tensor.

To illustrate this approach numerically, assume that the indicator of the fourth order system, specified by the A-matrix of (13), is defined by a coefficient *matrix* C:

$$\square \leftarrow C \leftarrow C1$$

⁻0.5	0	0.5	0
1	0	0	0

$$\rho C$$
$$2 \quad 4$$

$$(42)$$

Introducing this indicator of the output measurements together with the Cartesian tensor Q of (14) into the expression (34), produces the observability tensor:

$$\square \leftarrow M \leftarrow C + . \times Q$$

⁻0.5	0	0.5	0
0	⁻0.5	0	0.5
⁻2.5	1	2.5	0
0	⁻4.5	0	2.5
1	0	0	0
0	1	0	0
0	⁻2	0	0
0	4	0	0

$$(43)$$

$$\rho M$$
$$2 \quad 4 \quad 4$$

the dimensions of which are particularized by (35).

Applying (36) produces a 4-dimensional component array with the dimensions specified by (37):

$$\rho (\mathbb{Q} M) + . \times M$$
$$4 \quad 4 \quad 4 \quad 4$$

Redefining M by submitting the product (36) to a tensor contraction on the repeated index P, furnishes the symmetrical 2-dimensional array:

$$\square \leftarrow M \leftarrow + / [3] 1 \ 3 \ 3 \ 2 \mathbb{Q} (\mathbb{Q} M) + . \times M$$

7.5	⁻2.5	⁻6.5	0
⁻2.5	42.5	2.5	⁻11.5
⁻6.5	2.5	6.5	0
0	⁻11.5	0	6.5

$$(44)$$

$$\rho M$$
$$4 \quad 4$$

with the dimensions identified by (39).

Finally, submitting the thus specified square matrix M to the rank test of (40):

$$\rho \boxplus M$$
$$4 \quad 4$$

$$(45)$$

we obtain that the system is completely observable.

To round off this illustration of the testing for complete observability let us consider the situation whereby the system, represented by the Cartesian tensor Q of (14), is submitted to an indication of output measurements defined by a coefficient *vector* C. In particular, let this C-vector be the first row of (42):

$$\square \leftarrow C \leftarrow C2 \leftarrow C[1;]$$
$$\text{⁻0.5} \quad 0 \quad 0.5 \quad 0$$

$$(46)$$

Applying (34) results in a square matrix (with (41) defining its dimensions) representing the component array of the observability tensor:

$$\square\leftarrow M\leftarrow C+.\times Q$$

$$\begin{array}{cccc}
\bar{}0.5 & 0 & 0.5 & 0 \\
0 & \bar{}0.5 & 0 & 0.5 \\
\bar{}2.5 & 1 & 2.5 & 0 \\
0 & \bar{}4.5 & 0 & 2.5 \\
\end{array} \qquad (47)$$

$$\rho M$$

$$4 \quad 4$$

Performing the rank test (40) directly on this matrix:

$$\rho \boxminus M$$
$$\textit{DOMAIN ERROR} \qquad (48)$$
$$\rho \boxminus M$$
$$\wedge$$

tells us that the system is *not* completely observable.

2.5 Summing Up and Looking Ahead

The aim of the tensorial reformulation of the dual procedures for testing a proposed system configuration for the properties of completely state controllability and observability, is not actually that of facilitating the computations. Rather, the purpose here is to separate theoretically in the formulation as distinct physical properties, *the invariant structures of the three basic constituents: the system itself; its control; and its indicator for output measurements.*

Consulting Table 1, which for ready reference summarizes in its two columns all the pertinent facts of the dual tensorial formulations, the viewpoints, leading to a reevaluation of the conventional matrix approaches, may be explained as follows.

STATE CONTROLLABILITY	OBSERVABILITY
$Q\leftarrow A0,[2]A1,[2]....,[1.5]AP$ $(\underline{X},\underline{P},\underline{X})=\rho Q$	
$(\underline{X},\underline{U})=\rho B$	$(\underline{Y},\underline{X})=\rho C$
$K\leftarrow Q+.\times B$ $(\underline{X},\underline{P},\underline{U})=\rho K$	$M\leftarrow C+.\times Q$ $(\underline{Y},\underline{P},\underline{X})=\rho M$
$(\underline{X},\underline{P},\underline{P},\underline{U})=\rho K+.\times \lozenge K$ $K\leftarrow+/[3]1 \ 3 \ 3 \ 2\lozenge K+.\times \lozenge K$ $(\underline{X},\underline{U})=\rho K$	$(\underline{X},\underline{P},\underline{P},\underline{X})=\rho(\lozenge M)+.\times M$ $M\leftarrow+/[3]1 \ 3 \ 3 \ 2\lozenge(\lozenge M)+.\times M$ $(\underline{X},\underline{X})=\rho M$
$\rho \boxminus K$	$\rho \boxminus M$

Table 1. The Cartesian Tensor Approach

The fundamental duality between state controllability and observability is primarily a matter of a dual manner of interaction between the system and, respectively, its control and its output indicator. As listed in Table 1, the invariant system properties are defined by the Cartesian tensor Q while the control and the indicator are specified respectively by the Cartesian tensor sub-components B and C. Geometrically, we may conceive the latter as transformations to which we submit the system properties represented by the tensor Q. That is, the state controllability tensor K and the observability tensor M depict the system in two spaces as seen respectively, so to speak, from the input variables and the output variables. The causal relationship characterizing the interaction between the inputs and the system exhibits, by its very nature, a *directional duality* to that of the interaction between the system and its outputs. It is this directional duality that is reflected algebraically in the reversed generalized transpositions which enter the fundamental definitions in Table 1 of tensors K and M. Indeed, the fact that the rank tests proceed in identical ways, apart from the influence of the reversed generalized transpositions, indicates that *any geometrical duality beyond that of directional duality is purely accidental.* For example, we should not expect in general that because complete state controllability results from controlling a certain subset of state variables, then observing this subset or its disjoint dual will guarantee complete observability. In this connection, of course, the identical formulations of the rank tests are merely indications that will be substantiated later on from a purely structural point of view. In fact, the rank test is simply an algebraic procedure applied to verify whether tensors K or M are defined in a space which encompasses all state variables or only some subset.

Summing up, we may say that in comparison with the conventional matrix approach the tensorial formulation of state controllability and observability has brought three things explicitly to the fore. First, by depicting in terms of the separate tensor components Q, B, and C the properties of the system, its control, and its indicator, it preserved the distinct and invariant nature of the three basic components of the total configuration. Secondly, by representing in terms of basic tensor operations the interactions between the system, its control, and its indicator, the properties of state controllability and of observability respectively are derived from either of the two tensors K (for control) and M for measurement). Thirdly, by introducing reversed generalized transpositions in the definitions of tensors K and M, it is suggested that the structural relationship between state controllability and observability is a directional duality.

Based on these facts we would expect to find that, underlying the Cartesian tensor representation, there must be hidden fundamental topological relationships which characterize the basic structures of a proposed total system configuration. To be sure, such structural properties will determine only in a qualitative manner the potential of a proposed configuration with respect to state controllability and observability. That is, on the basis of a determination of the structural properties we can immediately reject proposals that have no possibilities of ever meeting the required specifications. Alternatively, if we consider a configuration that possesses the structural properties necessary to meet the specifications potentially, this is no guarantee that it will actually satisfy these requirements. Indeed, this can be verified only by a quantitative rank test for each choice of the coefficient values. Accordingly, determination of the structural properties of a configuration may yield qualitative design rules which, together with a knowledge of the physics of the problem and the costs of the different components, may guide us in the design process. For example, we may establish a priority list of components according to increasing costs and then, guided by physical feasibility, we may converge to a configuration the structural properties of which will be potentially satisfactory (of course, this includes the term rank test which we have assumed satisfied here). Now, trying out different gains and other quantitative values for this configuration, applying for each set of values the quantitative rank test, a final design may be accomplished. Thus, the development of a procedure for determination of the structural properties will naturally *subdivide the design procedure into two phases: a structural or topological phase and a quantitative or algebraic phase.*

However, these consequences for the design process will come true only, if two basic requirements are satisfied by our procedure for determination of the structural properties of a configuration. First, considered as qualitative design rules, it should be possible to represent the structural properties graphically in a simple manner so as to give a physical "feel" of the design problem. Secondly, it should be possible to determine the structural properties either by visual inspection of some diagram or by intuitively understandable computations. The development of a procedure that achieves this goal, is the aim of the following exposition.

Chapter 3. THE BOOLEAN TENSOR FORMULATION

This chapter will introduce a Boolean tensor formulation for the description of the structural properties of state controllability and observability. Essentially, this approach mirrors the Cartesian tensor formulation, but with all algebraic operations substituted by Boolean connectives. It is based on a group-theoretical consideration of the scale-forms of measurements which, applying a logical proposition to the entries of the coefficient matrices A, B, and C, permits us to substitute these entries by the truth-values 1 or 0 indicating whether or not the corresponding entry is non-zero. The outcome of the Boolean tensor approach, which amounts to a calculus of logical relations, is a set of topological properties that represents structurally the potential of a configuration with respect to state controllability and observability. From this arise the dual concepts of potential controllability and observability as the properties of a derived pair of Boolean tensors, the potential or structural controllability and observability tensors K and M. The presentation will be kept at a purely operational level with brief explanation only, since it exactly follows a general approach using APL which has been given in detail elsewhere (Franksen, 1976; and 1977).

3.1 Representing Structures by Boolean Tensors

Klein's Erlanger Program from 1872 gave coordinate geometry an entirely new perspective which fundamentally changed the theoretical basis of those empirical sciences, in particular physics, that relied on mathematics for abstract structures of modelling. His basic idea was that each geometry can be characterized by a mathematical group of transformations and that a geometry is really concerned with invariants under this group. Moreover, because some groups contain others as subgroups, some geometries will embrace others. That is, a subgroup of the transformation group of some geometry will define a subgeometry such that all theorems of the original geometry continue to be theorems in the subgeometry. Therefore, in an embracing sequence of geometries we may classify a theorem according to the geometry of the sequence in which it originates. Indeed, for any more general concept thus classified we may discover how successive subgeometries in the sequence make the concept gradually more specific and narrow by the addition of constraining details.

The purpose here is to go the opposite way. In other words, by removing the constraining quantitative details we wish to arrive at the larger and more general structural concepts. Evidently, this requires that we are able to ascertain the embracing sequence of geometries. As it turns out, the decisive invariant physical entity which permits us to establish the defining sequence of embracing subgroups, is the concept of *scale-form*. However, to evade further discussion of this subject let it suffice to say that whenever we relate by an *equivalence or order relation* measurements that may be quantitatively operated on also, we apply scale-forms defined by an embracing sequence of subgroups (Franksen, 1975). Intuitively, if we integrate the state equation (1) each row $A[I;]$ of the coefficient matrix A will identify by its non-zero entries a functional relationship that defines the corresponding state-variable $X[I]$ in terms of other state-variables possibly including itself. Consequently for the total system the non-zero entries of the A-matrix will depict a functional relation on the set of state-variables that reflects the structural properties of the system. To bring out the structure defined by this relation we simply consider all the entries of the A-matrix as quantitative measurements in the above sense of the intrinsic system properties. Now, to provide a measure on the most comprehensive and general scale-form of the thus identified structure we simply submit the entries of A to the logical proposition: *Assign the truth-value 1 to all measurements different from zero and the truth-value 0 to all measurements equal to zero.* The outcome of this process is a logical representation of the structural content of the functional relation defined on the state-variables by matrix A. In other words, we may conceive the truth-values as "measurements", on a more general scale-form, of the inherent physical structure. Submitting these logical measurements to Boolean operations, isomorphic to the real number operations on the corresponding quantitative measurements, will permit us to deduce the logical consequences that follows from the given structural properties. Hence, by introducing a Boolean tensor formulation, isomorphic to the Cartesian tensor representation, we arrive at a logical or qualitative description of the structural properties that comprises the total potential of any given physical configuration (Franksen, 1976). Evidently, emulating this approach for all four of the coefficient matrices in the state and output equations (1) and (2), the total set of structural properties may be established for the entire system.

A first step in this process is to apply the logical proposition to the coefficient matrix A:

$$A \leftarrow A \neq 0 \tag{49}$$

Accordingly, by substituting each non-zero entry in the coefficient matrix A by a 1, we have redefined A into a Boolean matrix of dimensions (X,X). To derive from this matrix, in a form corresponding to that of the Cartesian tensor Q (12), the structural properties of the system we must introduce the J'th Boolean power of matrix A isomorphic to the conventional power of matrix A (8). This forces us to consider the question of defining the Boolean matrix operation isomorphic to the conventional matrix product. To resolve this problem let us digress for a moment into the realm of mathematical notation.

Suppose that S and T are two matrices, conforming by a common dimension N, so that we can determine their matrix product $P=S\times T$. Then each element P_{ij} of the resulting matrix is determined by the compounded set of elementary scalar operations:

$$P_{ij} = \sum_{k=1}^{k=N} S_{ik} \times T_{kj} \tag{50}$$

Evidently, two basic operations are involved: a summation and a scalar product. In APL these two operations are reflected canonically in the notation of the compound operation $\ddot{+}.\times\ddot{}$ in terms of which the matrix product P is defined:

$$P \leftarrow S +.\times T \tag{51}$$

The advantage of this notation is not just that it is mnemonic and explicit. Far more important is it that it invites extensions to other operator combinations than + and × in a systematic and self-explanatory manner.

In Boolean algebra the summation $\ddot{+}\ddot{}$ is replaced by the logical connective *disjunction* $\ddot{\vee}\ddot{}$ (also called *or*), while the product $\ddot{\times}\ddot{}$ is substituted by the logical connective *conjunction* $\ddot{\wedge}\ddot{}$ (also called *and*). It follows from (51) that in APL the logical matrix product P of two conforming Boolean matrices S and T is written:

$$P \leftarrow S \vee.\wedge T \tag{52}$$

Computationally, the meaning of this expression is defined by the following isomorphism to (50):

$$P_{ij} = \bigvee_{k=1}^{k=N} S_{ik} \wedge T_{kj} \tag{53}$$

So much for the notational and formal side of the problem.[*]) As yet, we have not answered the real question; whether it is acceptable to introduce for Boolean matrices the inner product `"∨.∧"`, as the counterpart of the inner product `"+.×"` of conventional matrices. In symbolic logic or calculus of relations the inner product $S∨.∧T$ of two Boolean matrices S and T is known as the *relative product*. Thus, considering S and T as two relations their relative product is defined as the relation which holds between i and j there is an intermediate term k such that i has the relation S to k and k has the relation T to j. To illustrate, the relative product of *brother* and *father* is *paternal uncle*, and the relative product of *father* and *father* is *paternal grandfather*. It follows that (52) specifies P as the relation which results from the relative product of relations S and T.

Therefore, if we define isomorphic to (8) the J'th power of the square Boolean matrix A, denoted AJ, by the J relative products:

$$AJ←A∨.∧A∨.∧A∨.∧....∨.∧A \qquad (54)$$

the meaning of this expression is simply the logical consequences of applying to the relation A the transitive law J times. In other words, the process of making logical deductions from a relation A has been turned into a straight-forward calculation of non-negative and integral powers of A. Obviously, to make this process consistent, we define the 0'th power of A, designated $A0$, to be a unit matrix.

A valuable consequence of using the APL notation, is the fact that also the defined function for calculation of the J'th power of a Boolean matrix A may be established by an isomorphism to the function `"PLUSΔMPLY"` introduced previously:

```
        ∇ ORΔAND [□] ∇
      ∇ R←M ORΔAND I
[1]   ⍝
[2]   ⍝      THE NON-NEGATIVE AND INTEGRAL POWER
[3]   ⍝      "I" OF A SQUARE BOOLEAN MATRIX "M".
[4]   ⍝
[5]     →(I=0)/END
[6]     R←M∨.∧M ORΔAND I-1
[7]     →0
[8]   END:R←(⍳1↑⍴M)∘.=⍳1↑⍴M
      ∇
```

Hence, in either APL origin the J'th power of the Boolean matrix A is found in general by:

$$AJ← A \ \underline{ORΔAND} \ J \qquad (55)$$

where the dimensions of the different powers of A equal those of A itself (10). In this connection the reference to (10) is noteworthy, because it serves as a reminder that many operations on arrays are dependent only upon the shape of the arrays, but not upon the nature of their elements. Thus, by (12) we may define the component array of a Boolean tensor Q of rank 3 with its dimensions specified by (11). Evidently, this tensor Q is a Boolean polynomial or power series that depict the structural properties of the system as an invariant topological entity. For this reason we shall designate Q the *Boolean tensor polynomial*.

---

[*]) The consistency and formal elegance of APL as an explicit mathematical notation, may be illustrated by comparison of (52) with its counterpart expression introduced by Glover and Silverman (1976) in their section IV.

To illustrate numerically the establishment of Q consider the coefficient matrix A defined by (13). By (49) we assign truth values to the entries of A according to the relation of being non-zero:

$$\square \leftarrow A \leftarrow A \neq 0$$

$$\begin{array}{cccc} 0 & 1 & 0 & 0 \\ 0 & 1 & 0 & 0 \\ 0 & 0 & 0 & 1 \\ 1 & 0 & 1 & 0 \end{array}$$

(56)

Setting the number of powers $\underline{P}$ equal to the number of state-variables $\underline{X}=4$, implies that by (55) the following powers of the Boolean matrix A must be calculated:

$$\square \leftarrow A0 \leftarrow A \ \underline{OR \triangle AND} \ 0$$

$$\begin{array}{cccc} 1 & 0 & 0 & 0 \\ 0 & 1 & 0 & 0 \\ 0 & 0 & 1 & 0 \\ 0 & 0 & 0 & 1 \end{array}$$

$$\square \leftarrow A1 \leftarrow A$$

$$\begin{array}{cccc} 0 & 1 & 0 & 0 \\ 0 & 1 & 0 & 0 \\ 0 & 0 & 0 & 1 \\ 1 & 0 & 1 & 0 \end{array}$$

$$\square \leftarrow A2 \leftarrow A \ \underline{OR \triangle AND} \ 2$$

$$\begin{array}{cccc} 0 & 1 & 0 & 0 \\ 0 & 1 & 0 & 0 \\ 1 & 0 & 1 & 0 \\ 0 & 1 & 0 & 1 \end{array}$$

$$\square \leftarrow A3 \leftarrow A \ \underline{OR \triangle AND} \ 3$$

$$\begin{array}{cccc} 0 & 1 & 0 & 0 \\ 0 & 1 & 0 & 0 \\ 0 & 1 & 0 & 1 \\ 1 & 1 & 1 & 0 \end{array}$$

Introducing these results in (12) produces the Boolean tensor polynomial Q:

$$Q \leftarrow A0,[2]A1,[2]A2,[1.5]A3$$
$$\rho Q$$
$$4 \quad 4 \quad 4$$

(57a)

the dimensions of which, ρQ, are clarified by (11). In this connection, incidentally, the fact that the shape of the array is a cube, may be related mnemonically with the letter Q. Performing a generalized transposition of Q we may print it as follows for immediate verification:

$$
\begin{array}{cccc}
 & 2 & 1 & 3 \\
\end{array}Q
$$

$$
\begin{array}{cccc}
1 & 0 & 0 & 0 \\
0 & 1 & 0 & 0 \\
0 & 0 & 1 & 0 \\
0 & 0 & 0 & 1 \\
\\
0 & 1 & 0 & 0 \\
0 & 1 & 0 & 0 \\
0 & 0 & 0 & 1 \\
1 & 0 & 1 & 0 \\
\\
0 & 1 & 0 & 0 \\
0 & 1 & 0 & 0 \\
1 & 0 & 1 & 0 \\
0 & 1 & 0 & 1 \\
\\
0 & 1 & 0 & 0 \\
0 & 1 & 0 & 0 \\
0 & 1 & 0 & 1 \\
1 & 1 & 1 & 0 \\
\end{array}
$$

(57b)

3.2 The Concept of Potential State Controllability

The object now is to lay bare the fundamental structure underlying the quantitative or algebraic concept of state controllability. To achieve this we must reformulate the Cartesian tensor approach into a procedure based on corresponding Boolean tensors. A first indisputable step in this direction is to represent the structure of the control by a Boolean array B of rank 2:

$$B \leftarrow B \neq 0 \tag{58}$$

the dimensions of which are identified by (15).

With the Boolean conception of Q and B depicting the structural properties of the system and of the control respectively, we may then as in (16) establish an invariant topological entity by their structural combination. In this way a Boolean tensor K of rank 3 and with its dimensions defined by (17), is introduced. For obvious reasons it is called the *potential* or *structural state controllability tensor:*

$$K \leftarrow Q \vee . \wedge B \tag{59}$$

At this point, as a counterpart to the rank determination test, we have the problem of defining a corresponding set of operations on the Boolean tensor K. The purpose of the rank test is to verify that the Cartesian tensor Q has non-zero components relating each of the state variables X to at least one of the control variables U. Considering the geometrical image of K to the left of Fig. 4 we realize that, for each of the control variables U, we may investigate whether it is related to any given of the state variables X simply by performing a Boolean summation or reduction by the connective disjunction along dimension $\underline{P}$ of K. That is, to consider the total set of control variables U, assuming 1-origin, we must apply the socalled *existential quantifier* "∨/...." to the dimension $\underline{P}$ of K (Franksen, 1978):

$$\vee/[2]K \tag{60}$$

yielding a matrix of dimensions:

$$(\underline{X}, \underline{U}) = \rho(\vee/[2]K) \tag{61}$$

It is expected, for structurally complicated systems, that this matrix (60) may serve as a useful design aid because, in each column, it lists by 1's those state variables that are controlled by the control variable in question. Simultaneously, as also pointed out by Glover and Silverman (1976), a *zero row* in this matrix will signify a state variable that is *not* topologically or structurally related to any of the control variables. Accordingly, in such cases it will *never be possible* to control that state variable by the given set of control variables. If, alternatively, no zero row appears in this matrix, assuming the term rank test satisfied, the topological structure of the total system configuration with its control has the *potential* for complete state controllability. Whether or not this potential is actually used to attain this property, depends upon the quantitative values selected for the non-zero entries in A and B. To decide upon that matter the rank test must be executed. This, in essence is the concept of *potential state controllability*.

If we are not interested in dealing explicitly with the details of the Boolean design matrix (60), we may introduce, as a structural counterpart to (20), the Boolean sum or union of its columns:

$$K \leftarrow \vee / \vee / [2] K \qquad (62)$$

which is a vector of dimensions $\underline{X}$:

$$(\underline{X}) = \rho K \qquad (63)$$

The condition for potential state controllability of a system configuration with any given control, is that K defined by (62) is a full unit vector. In other words, it is required that the sum of the elements of K equals the number $\underline{X}$ of state variables. Hence, corresponding to the rank test (22), the test for potential state controllability may be formulated:

$$+ / K \qquad (64)$$

The advantage of performing the test in this manner, is that we obtain quantitative information on the number of state variables that satisfies the criteria even if the system itself is not potentially state controllable. However, to be formally consistent we should submit the vector K of (62) to the *universal quantifier* "$\wedge / \ldots$" (Franksen, 1978). In other words, a logical proposition which may be applied instead, is $\wedge / K$ that will return a 1 or a 0 depending upon whether or not potential state controllability is attained. *)

As usual, a special case will occur whenever B is a vector of dimension X. Clearly, in this situation the dimensions of the potential state controllability tensor K of (59) are given by (23). Therefore, we must replace (60) and (62) by the expression:

$$K \leftarrow \vee / K \qquad (65)$$

yielding a vector with its dimensions specified by (63). This vector, evidently, is then submitted to the test (64).

---

*) The superiority of APL as an explicit operational notation, may again be illustrated by comparing its simple and elegant implementation of the two logical quantifiers with the notational difficulties that confront Glover and Silverman (1976) in their description of the same test.

For comparison, let us apply the Boolean tensor approach to the numerical examples introduced to illustrate determination of state controllability by the Cartesian tensor procedure. Thus, corresponding to the *matrix* control defined by (24):

$$B1$$

```
1  0
0  2
0  0
0  0
```

we use (58) to define B:

$$\square \leftarrow B \leftarrow B1 \neq 0$$

```
1  0
0  1
0  0
0  0
```

$$\rho B$$

```
4  2
```

(66)

Applying (59) we find isomorphic to (25) that the potential state controllability tensor Q is:

$$\square \leftarrow K \leftarrow Q \vee . \wedge B$$

```
1  0
0  1
0  1
0  1

0  1
0  1
0  1
0  1

0  0
0  0
1  0
0  1

0  0
1  0
0  1
1  1
```

$$\rho K$$

```
4  4  2
```

(67)

Therefore, by (60) the Boolean design matrix is:

$$\vee / [2] K$$

```
1  1
0  1
1  1
1  1
```

(68)

which reveals, by its first column, that all state variables except $X[2]$ are potentially controlled by the first control variable $U[1]$ and, by its last column, that all state variables with no exception are potentially controlled by the second control variable $U[2]$. Further, since no zero rows appear the system is potentially state controllable. It is easy to verify, redefining K by (62):

$$\square \leftarrow K \leftarrow \vee / \vee / [2] K$$

```
1  1  1  1
```

(69)

and applying, corresponding to (27), the test (64):

$$+/K$$
$$4$$

$$(70)$$

that this is indeed the case. Of course, this is entirely in agreement with the fact that the total configuration is completely state controllable.

In the second example we submitted the system to a *vector* control specified by (30):

$$B2$$
$$0 \quad 0 \quad 3 \quad 0$$

which by (58) defines B as follows:

$$\square \leftarrow B \leftarrow B2 \neq 0$$
$$0 \quad 0 \quad 1 \quad 0$$
$$\rho B$$
$$4$$

$$(71)$$

By (59), corresponding to (31), the potential state controllability tensor Q is determined:

$$\square \leftarrow K \leftarrow Q \vee . \wedge B$$
$$0 \quad 0 \quad 0 \quad 0$$
$$0 \quad 0 \quad 0 \quad 0$$
$$1 \quad 0 \quad 1 \quad 0$$
$$0 \quad 1 \quad 0 \quad 1$$
$$\rho K$$
$$4 \quad 4$$

$$(72)$$

To obtain the Boolean design vector we apply (65):

$$\square \leftarrow K \leftarrow \vee /K$$
$$0 \quad 0 \quad 1 \quad 1$$

$$(73)$$

uncovering that only two of the state variables, $X[3]$ and $X[4]$, are controlled by the single control variable U. It follows, as may be verified by executing (64):

$$+/K$$
$$2$$

$$(74)$$

that the system is *not* potentially state controllable. Consequently, when we found by (32) that the configuration was not completely state controllable, this derives solely from its structural properties.

3.3 The Concept of Potential Observability

The essence of the concept of potential state controllability, is that, tensorially, it identifies the structural content of the causal relationship from control or input variables U to state variables X. Thus, the direction from variables U to variables X is defined topologically by this concept. The structural properties of the causal relationship from state variables X to output variables Y define, relative to variables X, a converse concept called potential observability. That is, whereas potential state controllability is based on a direction *into* variables X, potential observability is founded on the reverse direction *out from* variables X. Topologically, therefore, the two concepts are *directional duals* of each other. Based on this duality in formulation with the concept of potential state controllability the structural properties, defining the concept of potential observability, may be established as follows in terms of Boolean tensors.

First, corresponding to (58) we represent the structure of the indicator by a Boolean array C of rank 2:

$$C \leftarrow C \neq 0 \tag{75}$$

with the dimensions defined by (33).

Secondly, corresponding to (34) we establish as a dual to (59) a Boolean tensor M of rank 3 with its dimensions given by (35). This tensor is called the *potential* or *structural observability tensor*:

$$M \leftarrow C \vee . \wedge Q \tag{76}$$

Next, assuming 1-origin, we introduce corresponding to (60) the Boolean design matrix:

$$\vee / [2] M \tag{77}$$

with its dimensions specified by:

$$(\underline{Y}, \underline{X}) = \rho(\vee / [2] M) \tag{78}$$

The directional dual properties of (77) relative to (60) are expressed formally in terms of transposed relationships. Thus, each row of the matrix (77) lists by 1's those state variables that affect the output measurement in question. It follows that a *zero column* in this matrix will signify a state variable that is *not* topologically or structurally related to any of the output variables. Accordingly, in such cases it will *never be possible* to observe that state variable by the given set of output variables. If, alternatively, no zero column appears in this matrix the topological structure of the total system configuration with its output indicator, assuming the term rank test satisfied, has the *potential* for complete observability. Whether or not this potential is actually used to attain this property, depends upon the choice of quantitative values of the non-zero entries in A and C. To decide upon this matter a rank test must be executed. This, very briefly, is the concept of *potential observability*.

If we are not interested in dealing explicitly with the details of the Boolean design matrix (77), we may introduce, as a structural counterpart to (38), the dual operation of (62). That is, we perform the Boolean sum or union of the rows of (77):

$$M \leftarrow \vee / [1] \vee / [2] M \tag{79}$$

which is a vector of dimension X:

$$(\underline{X}) = \rho M \tag{80}$$

The condition for potential observability of a system configuration with any given indicator, is that M defined by (79) is a full unit vector. In other words, it is required that the sum of the elements of M equals the number $\underline{X}$ of state variables. Therefore, corresponding to (64), we replace the rank test (40) by the following test for potential observability:

$$+ / M \tag{81}$$

Of course, if we prefer to obtain the answer as a truth value, this test must be substituted by the logical proposition $\wedge / M$.

In the particular situation where C is a vector of dimension X, the dimensions of the potential observability tensor M of (76) are specified by (41). It follows that in such situations, corresponding to (65), we must substitute (77) and (79) by the expression:

$$M \leftarrow \vee / [1] M \tag{82}$$

yielding a vector with its dimension given by (80). It is this vector that we submit to the test (81) for potential observability.

As a numerical illustration of the test for potential observability let us re-examine those two examples which previously were submitted to the rank test for complete observability. Thus, corresponding to the *matrix* indicator defined by (42):

```
        C1
    ¯0.5         0         0.5         0
     1           0         0           0
```

we use (75) to define C:

```
             □←C←C1≠0
        1 0  1 0
        1 0  0 0
               ρC
          2 4
```
$$\tag{83}$$

Applying (76) we determine that the potential observability tensor M isomorphic to (43) is:

```
             □←M←C∨.∧Q
        1 0 1 0
        0 1 0 1
        1 1 1 0
        0 1 0 1

        1 0 0 0
        0 1 0 0
        0 1 0 0
        0 1 0 0
              ρM
         2 4 4
```
$$\tag{84}$$

By (77), therefore, the Boolean design matrix is:

```
          ∨/[2]M
        1 1 1 1
        1 1 0 0
```
$$\tag{85}$$

The first row of this matrix reveals that all four state variables X are observable by the first output variable $Y[1]$. The last row, on the other hand, shows that only the first two state variables, $X[1]$ and $X[2]$, are observable by the second output variable $Y[2]$. Further, since no zero columns appear the system is potentially observable. Redefining M by (79):

```
          □←M←∨/[1]∨/[2]M
        1 1 1 1
```
$$\tag{86}$$

and submitting this result to the test (81), corresponding to the rank test (45), we find:

```
          +/M
        4
```
$$\tag{87}$$

which tells us, in agreement with the fact that the total configuration is completely observable, that it is of course also potentially observable.

The second example is particularly interesting because here, for the first time, we are able to demonstrate the fundamental difference in viewpoint between the structural and the quantitative approaches in system design.

Assuming that the system is submitted to the *vector* indicator defined by (46):

$$\begin{array}{c} C2 \\ ^{-}0.5\ \ 0\ \ 0.5\ \ 0 \end{array}$$

we apply (75) to specify C:

$$\begin{array}{l} \square \leftarrow C \leftarrow C2 \neq 0 \\ 1\ \ 0\ \ 1\ \ 0 \\ \qquad \rho\, C \\ 4 \end{array} \tag{88}$$

By (76), corresponding to (47), the potential observability tensor M is determined:

$$\begin{array}{l} \square \leftarrow M \leftarrow C \vee . \wedge Q \\ 1\ \ 0\ \ 1\ \ 0 \\ 0\ \ 1\ \ 0\ \ 1 \\ 1\ \ 1\ \ 1\ \ 0 \\ 0\ \ 1\ \ 0\ \ 1 \\ \qquad \rho\, M \\ 4\ \ 4 \end{array} \tag{89}$$

Introducing this result into (82) we obtain the Boolean design vector:

$$\begin{array}{l} \square \leftarrow M \leftarrow \vee / [\,1\,] M \\ 1\ \ 1\ \ 1\ \ 1 \end{array} \tag{90}$$

It is evident, since this is a full unit vector, that the single output is causally related to all four of the state variables. Hence, as can be verified by executing the test (81):

$$\begin{array}{l} + / M \\ 4 \end{array} \tag{91}$$

the total configuration is potentially observable by virtue of the fact that it also satisfies the term rank test. But earlier, by failing to meet the rank test (48), we found that it was not simultaneously completely observable. Thus, we have here an illustration of a configuration that structurally possesses the potential for being completely observable, but which, by the actual choice of quantitative values, does not exploit the inherent possibilities of its topological properties.

The mere fact that we know that a configuration is potentially observable, permits us to look for simple revisions of the quantitative values. For example, by reversing the sign of first element of the C-vector in (46) we may attain the desired property:

$$\begin{array}{l} \square \leftarrow C \leftarrow C3 \\ 0.5\ \ 0\ \ 0.5\ \ 0 \end{array} \tag{92}$$

Thus, in this case by (34), assuming Q defined quantitatively by (14a), we obtain a Cartesian observability tensor of rank 2 the component array of which is the matrix:

$$\square \leftarrow M \leftarrow C + . \times Q$$

$$
\begin{array}{ccccc}
0.5 & & 0 & & \\
0 & & 0.5 & 0.5 & 0 \\
-2.5 & & -1 & 0 & 0.5 \\
0 & & -0.5 & 2.5 & 0 \\
& \rho M & & 0 & 2.5 \\
4 \quad 4 & & & &
\end{array}
\qquad (93)
$$

Submitting this matrix to the rank-test (40):

$$
\begin{array}{c}
\rho \boxminus M \\
4 \quad 4
\end{array}
\qquad (94)
$$

verifies that by this quantitative revision we have achieved our goal to make this configuration completely observable.

3.4 Summing Up and Looking Ahead

So far our aim has been to establish a Boolean tensor approach for determination of the structural or topological properties underlying the quantitative or alge- braic properties defining state controllability and observability of a total con- figuration. This gave rise to two new design concepts: potential state control- lability and potential observability which described topologically the structu- ral combinations of the system and its control respectively the system and its indication for output measurements. For ready reference and for comparison with the Cartesian tensor approach of Table 1 all the pertinent facts of the Boolean tensor formulation have been summarized in Table 2. In the last row of this table we have stated the test on quantitative form for purely pragmatic reasons. Of course, to be consistent with the mathematical logic we should substitute the plus-reduction "+/...." with the universal quantifier "∧/....".

STATE CONTROLLABILITY	OBSERVABILITY
$Q \leftarrow A0,[2]A1,[2]....,[1.5]AP$ $(\underline{X},\underline{P},\underline{X})=\rho Q$	
$(\underline{X},\underline{U})=\rho B$	$(\underline{Y},\underline{X})=\rho C$
$K \leftarrow Q \lor . \land B$ $(\underline{X},\underline{P},\underline{U})=\rho K$	$M \leftarrow C \lor . \land Q$ $(\underline{Y},\underline{P},\underline{X})=\rho M$
$\lor/[2]K$ $(\underline{X},\underline{U})=\rho(\lor/[2]K)$ $K \leftarrow \lor/\lor/[2]K$ $(\underline{X})=\rho K$	$\lor/[2]M$ $(\underline{Y},\underline{X})=\rho(\lor/[2]M)$ $M \leftarrow \lor/[1]\lor/[2]M$ $(\underline{X})=\rho M$
$+/K$	$+/M$

Table 2. The Boolean Tensor Approach

From a computational point of view the Boolean tensor approach is more efficient and requires less computer storage space than the Cartesian tensor approach, because all calculations are logical operations on the integer truth values 1 and 0. A further contribution in the same direction is the fact that the quantitative rank test is substituted by straight-forward logical summations or reductions. However, from the viewpoint of establishing simple design procedures the isomorphism between the two approaches imposes upon the Boolean tensor approach a seemingly unnecessarily complicated formulation. Therefore, the problem to be solved is to recast the Boolean tensor formulation into a more simple form amenable to an intuitively obvious graphical representation.

This is the problem that we are going to undertake in part II. In fact, we shall demonstrate there that a graphical representation and explanation of these concepts by means of digraphs will turn them into simple qualitative design rules. Furthermore, we shall also show how, by an elementary reformulation of the Boolean tensor procedure, we may arrive at a simple interpretation of these concepts in terms of digraph properties that are defined by the immediately intelligible reachability matrix.

CONCLUSION

Two things, we believe, are particularly noteworthy in connection with the tensorial approach. The first is that we identify the few fundamental Cartesian tensors and establish from this basis a set of derived tensors that will exhibit in tensorial form the different properties we wish to investigate. Most important among the fundamental tensors is the tensor Q of rank 3 that we have introduced to describe the intrinsic properties of the system configuration. Indeed, considering the Cayley-Hamilton theorem, this is the tensor on which a tensorial formulation of the stability problem must be based. The other thing to note is that the Boolean tensors are derived from the Cartesian tensors invoking Klein's Erlanger Program (Franksen, 1976).

Lin (1974) as well as Shields and Pearson (1976) argue that the only system parameter values known precisely are the zero values fixed, say by absence of physical connections between system parts. Since all other values are only approximate this leads them to a search for an alternative set of structurally identical values that will make the system completely controllable. It follows that their mathematical arguments must be based on considerations of matrix ranks and eigenvalues. Now, in our opinion a group-theoretical approach is both simpler and more fundamental (Franksen, 1978). The said parameter values will be measured on ratio or interval scales implying that they will be invariant under the similarity or the affine group of transformation. Substituting the measurements by Boolean values they are taken to a nominal scale that makes them invariant under the symmetric group of transformations. But the former two groups are subgroups of the latter. Hence, all the theorems that we may derive for the measurements under the symmetric group will continue to be theorems under the similarity or affine group. This is the essence of our application of Klein's Erlanger Program. Thus, by this approach the tensorial properties of the structural design concepts, potential controllability and potential observability, have been fixed in the form of a single condition called the reachability test. A digraph interpretation of this test and the introduction of a supplementary non-tensorial condition called the term rank test, will be given in part II.

PART II: THE GRAPH-THEORETICAL APPROACH

INTRODUCTION

Previously, in part I the conventional matrix criteria for quantitative determination of state controllability and observability were recast into dual invariant forms in terms of component arrays of Cartesian tensors. By isomorphism a corresponding dual pair of qualitative design concepts, called potential controllability and potential observability, were developed in terms of component arrays of Boolean tensors. The aim of the present part II is to derive and elucidate the latter pair of structural design concepts from a graph-theoretical point of view.

Interpretation of the state equation coefficient matrix as a digraph, depicting system states as vertices, furnishes an alternative formulation of potential controllability and potential observability in terms of invariant reachability properties between digraph vertices. Based on this interpretation these qualitative design concepts may be derived from the system digraph either directly by visual inspection or by a simple set of corresponding Boolean matrix operations. The latter set is distinguished by being computationally simple, yet theoretically fundamental. In fact, these operations will be shown to establish a basic Boolean isomorphism to the wellknown quantitative criterion for state controllability and observability originally proposed by Gilbert as an alternative to the Kalman formulation.

By this isomorphism Gilbert's assumption of distinct eigenvalues is turned into a Boolean assumption on what, in combinatorial mathematics, is knows as the *term rank* of the coefficient matrices (Ore, 1962; Ford & Fulkerson, 1962; Ryser, 1963). As a matter of terminology this concept lately has been named also the *generic rank* (Shields & Pearson, 1976; Glover & Silverman, 1976). However, to conform with the mathematical literature we shall adhere to the former designation. Essentially, the meaning of the concept is the maximal rank that a matrix may attain by virtue of its non-zero entries. It follows, as is wellknown from a variety of applications, that the term rank may be determined by scanning the rows and columns of the corresponding Boolean matrix for the entries of a maximal permutation matrix (Ford & Fulkerson, 1962). Though, as pointed out by Oystein Ore, an effective alternative tool may be the graph-theoretic *alternating path method* introduced by Julius Petersen in his investigations on the existence of subgraphs (Petersen, 1891; Ore, 1962).

Tieing up the concept of term rank of a Boolean matrix with the idea of alternating paths in the corresponding bipartite digraph, places this concept on a graph-theoretical foundation in common with the reachability concept discussed above. To be sure, both kinds of properties may be determined by visual inspection of a digraph, but this is rather a matter of convenience than of fundamental importance. Indeed, what is of fundamental importance, is that we may now identify, distinguish between, and compare reachability and term rank as distinct geometrical representations of structural properties that are invariant under different mathematical groups. Thus, digraph reachability, by virtue of being a Boolean tensor property, is invariant under the symmetric group or the group of all permutations. In contradistinction, term rank is not a Boolean tensor property since alternating paths are invariant only under the alternating group which, being the group of all even permutations, is a subgroup of the symmetric group.

Representing potential controllability and observability in terms of digraph reachability properties, gives rise to a generally valid, qualitative analogy of Gilbert's quantitative criterion for state controllability and observability completely disregarding his assumption of distinct eigenvalues. Endowing potential controllability and observability with the additional property of term

rank in terms of alternating paths, extends this analogy by a structural re-
straint that is the qualitative counterpart of Gilbert's quantitative assumption
of distinct eigenvalues. By the graph-theoretical approach to be undertaken in
the following, therefore, the concepts of potential controllability and observability will come out as direct generalizations of the wellknown quantitative
concepts of state controllability and observability.

Chapter 4. THE REACHABILITY CRITERION

Interpretation of the state equation coefficient matrix as a digraph, depicting
state variables as vertices and their mutual causal relationships as directed
edges, furnishes a simple graphical representation of the tensorial properties
of the qualitative design concepts: potential state controllability and potential observability. That is, assuming the term rank test to be satisfied, these
design concepts may be explained in terms of invariant reachability properties
between digraph vertices. It follows, that the tensorial properties of the two
design concepts may be established either directly by visual inspection of the
system digraph or indirectly by Boolean matrix operations based on the socalled
reachability matrix. The latter approach turns out to be of particular interest
by virtue of its isomorphism to an alternative set of definitions of the quantitative design concepts of state controllability and observability. This set
of alternative definitions, originating with Gilbert, is based on the assumption
that the state equation coefficient matrix has distinct eigenvalues (Gilbert,
1963). Invoking Klein's Erlanger Program to take this assumption to the Boolean
domain it becomes evident why an additional term rank test must be introduced.

4.1 A Digraph Interpretation

Graph theory is a particular subject within combinatorial or algebraic topo-
logy which is concerned with continuity properties of geometric structures
disregarding distances and angles. Especially directed graphs, the socalled
digraphs, may be conceived as a Boolean analogy of coordinate geometry. In
other words, by virtue of the fact that the Boolean tensors of potential state
controllability and potential observability represent the inherent topological
properties of a total configuration, we may represent these topological structures by digraphs the "coordinates" of which, as we shall see, are defined by
the coefficient matrices of the state and output equations (I.1) and (I.2). *)

The purpose here is not to discuss the theory of digraphs, but merely to apply its wellknown results. So we shall adopt the Boolean matrix A of (I.49)
as the algebraic representation of the topological coordinates of the system structure. The manner in which we have formulated the Boolean tensor
approach, forces us to consider the logical matrix A as a *nodal successor matrix*
if, in the digraph, we wish to represent the state variables X each by a separate vertex. If I is a row index and J a column index of A an entry $1=A[I;J]$
signifies that state variable $X[I]$ is the *immediate successor* of state variable
$X[J]$. In the corresponding *successor digraph* this is represented by an oriented or directed edge from vertex J to vertex I. A special case of these rules
is an entry $1=A[I;I]$ which introduces in the digraph a directed edge, a socalled
loop, with the same vertex I as both initial and final vertex. As an illustration the successor digraph, representing the topological system structure defined by (I.56):

$$\square \leftarrow A \leftarrow A \neq 0$$

$$\begin{matrix} 0 & 1 & 0 & 0 \\ 0 & 1 & 0 & 0 \\ 0 & 0 & 0 & 1 \\ 1 & 0 & 1 & 0 \end{matrix} \tag{1}$$

has been established in Fig. 1. In this digraph a vertex number N identifies
the corresponding state variable $X[N]$.

---

*) A reference number preceded by the Roman numeral I, refers to the item of
that number in part I.

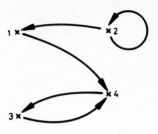

Fig. 1.   A Successor Digraph of State-Variables

Reachability, as used here, is a term applied to generally describe continuity properties of a digraph that depends upon the individual orientations of the directed edges. In particular, a vertex I *is reachable* from vertex J if it is possible to trace a path along the edges, always going in the direction of the arrows, from vertex J to vertex I. Thus, in Fig. 1 vertex 3 is reachable from vertex 2, but the opposite is not true. A path in which we always go in the direction of the arrows, is called *a directed path.* We measure the *topological distance* along any directed path by counting the number of directed edges that we have to follow. For example, in Fig. 1 the distance from vertex 2 to vertex 4 is 2, and from vertex 2 to vertex 3 it is 3. Incidentally, the last distance is a *minimum* distance since we may extend the distance of this path indefinitely by a continued cycling around between vertices 3 and 4. Clearly, reachability depicts a transitive binary relation on whatever entities are represented by a finite number of vertices. Since in practice we may always compare an entity by itself, we must assume that any vertex of a digraph is reachable from itself. This brings in the special question of defining a distance from a vertex to itself. Conventionally, we assign this distance a zero value based on the argument that a vertex is reachable from itself along a directed path involving a zero number of directed edges.

In general for any digraph a vertex I may be reachable from a vertex J along different directed paths and over different topological distances. For example, in Fig. 6 vertex 3 will be reachable from vertex 4 over the distances 1,3,5 etc. Similarly, vertex 2 will be reachable from itself over any distance from zero and upwards. This suggests that an organized way of investigating the reachability of a digraph will be to determine for each distance: 0,1,2,3...etc. the directed paths of that distance. In this connection it is evident that the term *immediate successor* refers to the reachability of a digraph along directed paths of distance 1. Algebraically, we represented the reachability over distance 1 by the Boolean A-matrix (I.49) considered as a nodal successor matrix. It is evident, as may be illustrated by comparing (I.57b) with Fig. 1, that the J'th power of the Boolean A-matrix, A^J, defined by (I.54), specifies the reachability of the digraph over all directed paths of distance J. In particular, this explains the meaning of the unit matrix A^0 defined by the zero'th power of A. Therefore, the Boolean tensor polynomial Q, defined by introducing (I.54) into (I.12), is indeed a basic topological invariant because, along its dimension P, it represents the reachability over directed paths of all possible distances.

At this point the distinction between the graph-theoretical concepts of reachability and successorship should be made absolutely clear. Reachability of one vertex I from another J, is the fundamental topological property that at least one directed path exists in the digraph from vertex J to vertex I. The concept of a successor is a prescription for uniquely denoting the reachability relationship between two vertices such that we may explicitly identify, say, the causal or physical meaning of this relationship relative to the entities represented by the pair of vertices. Consequently, the successor interpretation of the reachability of vertex I from vertex J, is that vertex I is a successor of vertex J. Obviously, maintaining the successor interpretation we may introduce the converse reachability relationship simply by reversing this statement. Generalizing this idea it is easy to see that the *directional dual of a digraph* or, more briefly, the *converse digraph* (i.e., the digraph with the directions of all edges reversed), is represented algebraically by the transpose of its nodal successor matrix. For example, in the same manner as the Boolean coefficient matrix A of (I.56) defined the digraph of Fig. 1, an immediate successor interpretation of its transpose:

$$
\phi A
$$

$$
\begin{array}{cccc}
0 & 0 & 0 & 1 \\
1 & 1 & 0 & 0 \\
0 & 0 & 0 & 1 \\
0 & 0 & 1 & 0
\end{array}
\tag{2}
$$

will define its converse digraph depicted in Fig. 2. To be sure, it was this relationship that was referred to previously in part I when, with reference to the reversed generalized transpositions in the definitions of the two tensors K and M, it was stated that *the structural relationship between state controllability and observability is a directional duality.*

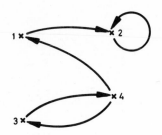

Fig. 2. A Converse or Predecessor Digraph

For the purpose of actually drawing the converse digraph of Fig. 2 it is not at all necessary first,either to draw the successor digraph of Fig. 1,or to produce the transpose of A. We may simply interprete the Boolean A-matrix of (1) as an *immediate predecessor* relationship. That is, we conceive an entry $1 = A[I;J]$ to mean that state variable $X[I]$ is the immediate predecessor of state variable $X[J]$ and enter a corresponding directed edge in the digraph from vertex I to vertex J. However, not to confuse the issue we shall assume arbitrarily in the following that all digraph interpretations are expressed exclusively in terms of successor relationships.

Now, to get a physical "feel" for the meaning of the concepts of potential state controllability and potential observability let us try to give the numerical results obtained previously a digraph interpretation. The system configuration in all of these examples is defined topologically by the Boolean A-matrix of (1). Interpreting this matrix as a nodal successor matrix we may depict the structure of the system by the digraph of Fig. 1 in which the vertex numbers refer to the state variables. The reachability properties over all distances of this digraph are represented algebraically by the Boolean tensor polynomial Q. The transpose of A provides us with the directional dual structure that is depicted by the converse digraph of Fig. 2. The generalized transpose of Q supplies us with the reachability properties over all distances of the converse digraph.

Topologically, assuming the term rank test satisfied, the concept of potential state controllability is concerned with the reachability properties of the system structure from those state variables, that are directly controlled by the input variables, to all the remaining state variables. In other words, our problem is to determine which of the remaining state variables that are successors of the directly controlled state variables. Evidently, the answer to this question may be obtained by visual inspection of the successor digraph of Fig. 1.

In the first numerical example we were concerned with a control the structure of which was defined by the Boolean matrix B of (I.66):

$$
\begin{array}{cc}
& B \\
1 & 0 \\
0 & 1 \\
0 & 0 \\
0 & 0 \\
& \rho B \\
4 & 2
\end{array}
$$

Obviously, as may be verified by (I.15), two input variables, controlling directly state variables $X[1]$ and $X[2]$ respectively, are involved. Marking off vertex 1 in the digraph of Fig. 1 it is immediately seen that its successors are vertices 3 and 4. In other words, controlling vertex 1 makes vertices 3 and 4 potentially controllable. Similarly, marking off vertex 2 we find that the remaining three vertices are all its successors. Hence, controlling vertex 2 makes all the remaining three vertices potentially controllable. Thus, a mere visual inspection of the system digraph provides us, in a qualitative and simple manner, with all the information needed for deciding whether or not the given control will make the system potentially state controllable. Of course, we have obtained this information previously in the two columns of the Boolean design matrix (I.68). However, the new thing is, that not only is it an easy and rapid process to draw a digraph, but, even more important, inspection of the digraph may result in a host of ideas on how to actually design a control that will meet the required specifications.

In the second numerical example the structure of the control was given by the Boolean vector B of (I.71):

$$
\begin{array}{c}
B \\
0 \quad 0 \quad 1 \quad 0 \\
\rho B \\
4
\end{array}
$$

Here only the state variable $X[3]$ is directly controlled by the single input variable. Marking off the corresponding vertex 3 of the digraph of Fig. 1 immediately reveals, in agreement with the Boolean design vector (I.73), that its only successor (apart from itself at distance 0) is vertex 4. That is, vertices 1 and 2 are not reachable from this control and the system is not state controllable by any control exhibiting this structure.

Now, still assuming the term rank test to be satisfied, let us turn to the digraph representation of the concept of potential observability. Topologically, this concept is related to reachability properties of the system structure going to those state variables that are directly measured as outputs from all of the remaining state variables. In other words, our problem is the converse of before, namely, that of determining which of the remaining state variables that have the directly measured state variables as successors. Obviously, we may turn this problem into the previous one by consulting the converse digraph of Fig. 2 for successors of the state variables directly measured.

The first numerical example is the indicator defined topologically by the Boolean C-matrix of (I.83).

$$
\begin{array}{c}
C \\
\begin{array}{cccc}
1 & 0 & 1 & 0 \\
1 & 0 & 0 & 0 \\
\end{array} \\
\rho C \\
2 \quad 4
\end{array}
$$

Evidently, as may be verified by (I.33), this indicator introduces two output variables. The first of these measures directly state variables $X[1]$ and $X[3]$, whereas the second measures directly $X[1]$. Visual inspection of Fig. 2 tells us that vertex 2 is the only successor of vertex 1, but that all the remaining three vertices: 4, 1, and 2 are successors of vertex 3. Hence, in agreement with the corresponding information obtained from the Boolean design matrix (I.85), the structure of this indicator makes the system potentially observable.

The second numerical example derives from the Boolean C-vector of (I.88):

$$
\begin{array}{c}
C \\
\begin{array}{cccc}
1 & 0 & 1 & 0 \\
\end{array} \\
\rho C \\
4
\end{array}
$$

Clearly, since the structure of this indicator defines a direct measurement of the same two state variables $X[1]$ and $X[3]$ the observations made previously, will apply. That is, since all vertices of the converse digraph of Fig. 2 are successors of vertex 3 we find, in agreement with the Boolean design vector (I.90), that potential observability will result.

4.2 Applying the Reachability Matrix

The digraph interpretation of the system structure made it clear that state controllability and observability are directional dual concepts that originate in the reachability properties of converse digraphs. Thus, the successor relationships between the vertices of the digraphs provided a simple picture illustrating the concepts of potential state controllability and potential observability and their use as qualitative design rules. However, for higher order systems it is likely to expect more complicated digraphs for which a visual inspection may be error prone. For this reason and also to elucidate the fundamental role as a structural tool that digraphs may play within the field of control theory, we wish to introduce some of the wellknown and rather simple procedures for computing the reachability properties of a digraph.

The basic topological concept in this respect is the socalled *reachability matrix* R that defines the reachability over all distances between the vertices of a digraph. Adopting the successor interpretation we may derive it from the Boolean tensor polynomial Q applying the existential quantifier along the dimension $\underline{P}$:

$$ R \leftarrow \vee / [2] Q \tag{3} $$

Clearly, in spite of its name conventionally adopted in the literature, we should consider R a Boolean tensor of rank 2 with its dimensions identified by:

$$ (\underline{X}, \underline{X}) = \rho R \tag{4} $$

It depicts, as an invariant entity in a very compressed form, the combined structural properties of reachability of a digraph. To illustrate this we may compare the reachability properties of the digraph of Fig. 1 by inspection with the reachability matrix R obtained by submitting Q of (I.57a) to (3):

$$\Box \leftarrow R \leftarrow \vee/[2]Q$$

$$\begin{array}{cccc} 1 & 1 & 0 & 0 \\ 0 & 1 & 0 & 0 \\ 1 & 1 & 1 & 1 \\ 1 & 1 & 1 & 1 \end{array}$$

$$\rho R$$
$$4 \quad 4$$

(5)

Now, for a digraph with a finite number X of vertices the longest directed path, making a vertex reachable from another, can at most contain (X-1) directed edges, since a directed path of minimum distance suffices to establish a reachability relationship between a pair of vertices. It is for this reason that the number P of powers involved of A, is given by (I.3) in the conventional matrix criteria for completely state controllability and observability. However, unless the digraph defines a single directed path going through all of its vertices all possible reachability relationships between the vertices may be established at a minimum distance less than P=X-1. In other words, to determine R we need only calculate that number of powers of A which is actually necessary relative to the digraph at hand. A straight-forward function, valid in either APL origin, that determines the reachability matrix R in accordance with this principle, is the following:

```
        ∇ REACHABILITY [□] ∇
      ∇ R←REACHABILITY M;U;AUX
[1]   ⍝
[2]   ⍝      THE REACHABILITY MATRIX "R" IS ESTABLISHED
[3]   ⍝      FROM A SQUARE BOOLEAN MATRIX "M".
[4]   ⍝
[5]   INITIAL:R←U←(⍳1↑⍴M)∘.=⍳1↑⍴M
[6]   NEXTΔPOWER:AUX←R
[7]   R←U∨M∨.∧R
[8]   →(∨/∨/AUX≠R)/NEXTΔPOWER
      ∇
```

Submitting the Boolean matrix A of (1) to this function produces, in agreement with (5), the following reachability matrix:

$$\Box \leftarrow R \leftarrow REACHABILITY \ A$$

$$\begin{array}{cccc} 1 & 1 & 0 & 0 \\ 0 & 1 & 0 & 0 \\ 1 & 1 & 1 & 1 \\ 1 & 1 & 1 & 1 \end{array}$$

$$\rho R$$
$$4 \quad 4$$

(6)

In practice faster and less storage-demanding computer methods exist for calculation of the reachability matrix.

To verify the results of a visual inspection of a digraph, representing a system structure combined with the structure of a control or an indicator, we must determine the corresponding Boolean design matrix or vector. To see how this is done directly in terms of the reachability matrix R, let us consider the topological combination of a system, specified by the Boolean tensor polynomial Q, and a control, identified by a Boolean matrix B. Clearly, in this case, by (I.60), the Boolean design matrix, derived from the potential state controllability tensor K, may be expressed:

$$\vee/[2](Q\vee.\wedge B) \tag{7}$$

From (I.17) we recall that the second dimension in this expression is the dimension P of powers of A. Hence, an alternative formulation of (7) is the expression:

$$(A0\vee.\wedge B)\vee(A1\vee.\wedge B)\vee....\vee(AP\vee.\wedge B) \tag{8}$$

but by the distributive law this may be rewritten:

$$(A0\vee A1\vee....\vee AP)\vee.\wedge B \tag{9}$$

where we recognize by comparison with (3) that:

$$R\leftarrow A0\vee A1\vee....\vee AP \tag{10}$$

Therefore, instead of determining the Boolean design matrix by (7), we may calculate it directly by the computationally much more simple, relative product:

$$R\vee.\wedge B \tag{11}$$

which, if we so wish, may be given the interpretation:

$$(\vee/[2]Q)\vee.\wedge B \tag{12}$$

To illustrate numerically this simplification in formulation and calculation consider the Boolean control matrix B specified by (I.66):

$$B$$
$$
\begin{array}{cc}
1 & 0 \\
0 & 1 \\
0 & 0 \\
0 & 0 \\
\end{array}
$$
$$\rho B$$
$$4 \quad 2$$

Implementing this control on the system produces, by (7) and in agreement with (I.68), the Boolean design matrix:

$$\vee/[2](Q\vee.\wedge B) \tag{13}$$
$$
\begin{array}{cc}
1 & 1 \\
0 & 1 \\
1 & 1 \\
1 & 1 \\
\end{array}
$$

Alternatively, the same result is directly obtained from the reachability matrix R by (11):

$$R\vee.\wedge B \tag{14}$$
$$
\begin{array}{cc}
1 & 1 \\
0 & 1 \\
1 & 1 \\
1 & 1 \\
\end{array}
$$

By consideration of duality we may transfer this approach from the concept of potential state controllability to that of potential observability. The result of this is that the Boolean tensor operations, summarized in Table I.2, may be substituted by the far more simple Boolean matrix operations, listed in Table 1, and derived from topological properties of the reachability matrix R.

The computational advantages of the reachability matrix approach given in Table 1 are two-fold. First, all operations are performed on matrices or vectors and at no point is it necessary to consider operations on 3-dimensional arrays. Thus, the reachability matrix R may be established iteratively from a minimum set of powers of A without ever explicitly considering the Boolean tensor polynomial Q. Secondly, the introduction of the reachability matrix in the formulation eliminates the necessity for determining the potential state controllability tensor K (I.59) and the potential observability tensor M (I.76). In other words, representing the system structure by the reachability matrix R instead of the Boolean tensor polynomial Q in the expressions for determination of K and M, the Boolean design matrices $\lor/[2]K$ (I.60) and $\lor/[2]M$ (I.77) are established directly.

STATE CONTROLLABILITY	OBSERVABILITY
$R\leftarrow\lor/[2]Q$ $(\underline{X},\underline{X})=\rho R$	
$(\underline{X},\underline{U})=\rho B$	$(\underline{Y},\underline{X})=\rho C$
$K\leftarrow R\lor.\land B$ $(\underline{X},\underline{U})=\rho K$	$M\leftarrow C\lor.\land R$ $(\underline{Y},\underline{X})=\rho M$
$K\leftarrow\lor/K$ $(\underline{X})=\rho K$	$M\leftarrow\lor/[1]M$ $(\underline{X})=\rho M$
$+/K$	$+/M$

Table 1. The Reachability Matrix Approach

In Table 1, to emphasize this isomorphism, we have introduced for these design matrices the designations K and M. Of course, as may be verified by a comparison of Tables I.2 and 1, the basic tests on the Boolean design matrices remain unaltered. In particular, assuming the term rank test satisfied, a system is *potentially state controllable* if the Boolean design matrix (11):

$$K\leftarrow R\lor.\land B \qquad (15)$$

has *no rows that are all zeros*. Under the same assumption, similarly, a system is *potentially observable* if the Boolean design matrix:

$$M\leftarrow C\lor.\land R \qquad (16)$$

has *no zero columns*. These two criteria, as explained in part I, lead to the test procedures listed in Table 1 if, for pragmatic reasons, we substitute the universal quantifier by the corresponding summation or plus reduction.

Summing up, the representation of the system structure by a digraph and the interpretation of the tensorial properties of potential state controllability and potential observability as reachability properties of that digraph, provide a simple graphical illustration that may serve as a qualitative design tool for automatic control systems. It is noteworthy in this connection that the actual design process, assuming the term rank test to be satisfied, may be guided either by a visual inspection of the digraph or by a simple set of corresponding Boolean computations on the reachability matrix.

4.3 A Basic Isomorphism

The formulations in terms of eqs. (15) and (16) of the reachability criterion for potential state controllability and potential observability, exhibit a fundamental isomorphism to the formulations of a quantitative criterion for state controllability and observability originally proposed by Gilbert as an alternative to the Kalman formulations summarized in chapter 2 section 2.1 (Gilbert, 1963; Kalman, 1963). The importance of this isomorphism, is that it explains the nature of the assumption of the term rank property on which the reachability criterion must be based. Of course, at the Boolean level the term rank test and the reachability test will appear as independent criteria. However, as the following discussion of this isomorphism will disclose, the term rank property derives from a fundamental assumption adopted by Gilbert in his alternative formulation of a quantitative criterion. It follows that, at least conceptually, the term rank condition will appear as a fundamental assumption underlying the reachability criterion. Incidentally, this agrees with the computational test procedure for potential controllability recommended by Glover and Silverman (1976) for practical reasons.

Considering the coefficient matrix A of the system state equation (I.1) Gilbert founded his alternative quantitative criterion on the basic assumption that *the eigenvalues of the state coefficient matrix A are all distinct*. The purpose of this assumption, to be sure, is to guarantee the existence of a diagonalizing transformation of the A-matrix that will bring the state equation (I.1) into its canonical form. Thus, by virtue of this assumption the eigenvectors are linearly independent. The implication hereof is that a non-singular matrix, say E, may be formed with the eigenvectors as columns. Hence, the dimensions of the *eigenvector matrix* E, defining the diagonalizing transformation, coincide with the dimensions of matrix A (I.10):

$$(\underline{X} , \underline{X}) = \rho E \qquad (17)$$

Now, submitting the vector X of state-variables to a linear transformation specified by the *inverse* of the eigenvector matrix E, produces a vector Z of new state-variables called the *normal coordinates* by Gilbert:

$$Z \leftarrow (\boxdiv E) + . \times X \qquad (18)$$

Premultiplying this expression by matrix E we obtain:

$$E + . \times Z = X \qquad (19)$$

which introduced into the state and output equations, (I.1) and (I.2), brings them into their canonical forms. Hence, temporarily adopting for comparison a conventional mathematical notation, the latter two equations may be rewritten:

$$\dot{Z} = L \times Z + K \times U \qquad (20)$$

$$Y = M \times Z + D \times U \qquad (21)$$

In the former expression (20), turning back to the APL notation, matrix L is the *diagonal* matrix of the distinct eigenvalues of A produced by the *similarity* transformation:

$$L \leftarrow (\boxminus E) + . \times A + . \times E \tag{22}$$

Thus, by this transformation the state-variables are completely decoupled from each other. It follows that the only way the state-variables are controllable, is through the inputs U directly. This gives rise to Gilbert's formulation that a system is *completely state controllable* if the derived quantitative matrix:

$$K \leftarrow (\boxminus E) + . \times B \tag{23}$$

has *no rows that are all zeros*. Similarly, by considering the expression (21), Gilbert arrived at the conclusion that a system is *completely observable* if the derived quantitative matrix:

$$M \leftarrow C + . \times E \tag{24}$$

has *no zero columns*.

Placing Gilbert's two quantitative conditions (23) and (24) side by side with the two qualitative conditions (15) and (16) respectively, unveils a remarkable identity in the abstract pattern underlying the two sets of conditions. Identity of mathematical form is bound to have some deeper meaning, which can be grasped as an idea rather than as a collection of symbols. It will be worthwhile, therefore, to investigate this structural property more closely. Clearly, this structural identity is an isomorphism, but with one important exception. In the comparison of (23) with (15) the reachability matrix corresponds to the inverse of the eigenvector matrix:

$$R \leftrightarrow \boxminus E \tag{25a}$$

whereas in the comparison of (24) with (16) the reachability matrix corresponds to the eigenvector matrix itself:

$$R \leftrightarrow E \tag{25b}$$

A first thought may be that the reachability matrix R is its own inverse. However, it has been shown that a Boolean inverse of a square Boolean matrix M will exist only if M is *orthogonal* (Luce, 1952). Hence, the Boolean inverse of M will be its transpose satisfying the isomorphic properties:

$$I = M \vee . \wedge \lozenge M \qquad \text{and} \qquad I = (\lozenge M) \vee . \wedge M \tag{26a\&b}$$

where I denotes the identity matrix. Now, in conventional matrix algebra the condition that a square matrix M has an inverse, is that M is *non-singular*. The corresponding condition that a square Boolean matrix M has a Boolean inverse, is that M is a *permutation matrix* (Luce, 1952; Franksen, 1976). In other words, *only a permutation matrix will have a Boolean inverse, namely its transpose*. It follows that the reachability matrix R cannot directly satisfy correspondences of the forms of hypothesized by (25).

Still, to solve this problem let us assume that the state and output equations have been transformed into their canonical forms (20) and (21). Thus, by (22) the state coefficient matrix A has been brought into the diagonal matrix L of distinct eigenvalues. It is wellknown that any arbitrary J'th power of A, denoted AJ, may be diagonalized by a similarity transformation defined, in complete analogy with (22), by the eigenvector matrix E:

$$LJ \leftarrow (\boxminus E) + . \times AJ + . \times E \tag{27}$$

since AJ has identically the same eigenvectors as A. But this implies that the Cartesian tensor Q (I.12) may be recast by the similarity transformation:

$$Q \leftarrow (\boxminus E) + . \times Q + . \times E \tag{28}$$

into the canonical form illustrated in Fig. 3 (see also Fig. I.2). The characteristic feature of this form, is that the non-zero entries of the distinct eigenvalues and their respective powers are concentrated in a diagonal plane through the dimension P of powers of A with zero entries elsewhere.

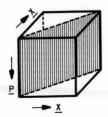

Fig. 3.   Cartesian Tensor Q on Diagonalized Form

It becomes obvious, at this point, that if the Boolean tensor polynomial Q is established from the state equation on the canonical form (20), then also the Boolean tensor Q will take the diagonalized canonical form exhibited in Fig. 3. By (3), accordingly, the reachability matrix R will be a diagonal Boolean matrix. Neglecting for the moment the possibility that one of the distinct eigenvalues may be zero, the diagonalized reachability matrix will appear in the form of the identity matrix I. Evidently, since now the reachability matrix is the only permutation matrix which by (26) is its own Boolean inverse, we see that the correspondence (25) holds. Thus, we may conclude that indeed *a fundamental isomorphism exists between Gilbert's quantitative criteria and the reachability criteria.*

But this brings us to the question of establishing the Boolean assumption corresponding to Gilbert's assumption of the distinct eigenvalues guaranteeing the existence of the non-singular eigenvector matrix E. To find this correspondence let us reexamine our approach in order to determine the structural significance of Gilbert's assumption.

4.4 Some Group Theoretical Implications

Clearly, the fundamental operation invoked by Gilbert's assumption is the *similarity* transformation (22) based on consideration of quantitative square matrices for which the invariant properties under the similarity group of transformations are rank, determinant, trace, and identical eigenvalues with the same characteristic multiplicities. However, by assuming distinct eigenvalues in order to ensure that the state coefficient matrix A is *diagonable* (i.e. it is similar to a diagonal matrix), he goes even further by considering a subgroup of the similarity group of transformations. Of course, by virtue of being a subgroup it will exhibit, as a subset, the invariant properties of the similarity group. Thus, the problem of determining the structural significance of Gilbert's assumption, is tantamount to a representation of the invariant properties of this subgroup in the Boolean domain. Evidently, these properties are related to the question of rank and the distinction between non-zero and zero eigenvalues also possibly touching upon their characteristic multiplicities.

At the outset, before we enter into a determination of the subgroup and its Boolean representation, let us glance at the meaning of our approach. Similarity transformations, conceived as space transformations, simply multiply all distances by a non-zero, usually positive, factor of proportionality. Thus, searching after those vectors X which are transformed into scalar multiples of themselves by some constant coefficient matrix A we arrive, adopting the conventional mathematical notation, at the wellknown equation:

$$A \times X = \lambda \times X \tag{29}$$

for determination of the eigenvalues λ. The solution vectors X, each associated with an eigenvalue, are, of course, the eigenvectors E fixing only the invariant directions in space by virtue of the fact that they have arbitrary length. In other words, if X is a solution vector then, for any non-zero scalar K, also K×X will be a solution vector. The invariance of this property under the group of similarity transformations is important because it defines the elements of vector *X* as measurements on a *ratio scale* (Stevens, 1946; Franksen, 1975). That is, the scale-form admits a change of unit, but the zero point must remain fixed.

Tieing up the concept of a ratio scale with linear transformations of the type (18) and (19), explains why any quantitative state coefficient matrix A must submit to a similarity transformation such as (22) but without necessarily leading to anything more than another square matrix. However, according to a well-known theorem, formulated in 1909 by I. Schur, there exists a *unitary* matrix E the similarity transformation by which will bring any A into a *triangular form* with the eigenvalues appearing on the diagonal (Bellman, 1960; Lancaster, 1969). The inverse of a unitary matrix E, we recall, is its complex conjugate transpose. Thus, if, as a special case, all the entries of E are real, the inverse is simply its transpose. In APL if we map each complex number into a 2×2 real matrix as follows:

$$a + jb \leftrightarrow \{ \begin{smallmatrix} a & b \\ -b & a \end{smallmatrix} \} \text{ where } j^2 = -1 \tag{30}$$

then the generalized transpose ⍉E will designate the transpose or the conjugate transpose of E accordingly as E represents a real or a complex array. Based on this idea the unitary similarity transformation involved in Schur's theorem may be written:

$$(⍉E)+.\times A+.\times E \tag{31}$$

A particularly interesting fact, is that if A is *normal* (i.e., A commutes with its conjugate transpose) then, by a transformation like (31), A is *unitarily similar* to the *diagonal* matrix of its, possibly multiple, eigenvalues.

Independent of the fact whether A is diagonable, the importance of the unitary similarity transformation is that, as a consequence of Schur's theorem, the rank of any quantitative square matrix A may be determined by considering its total number of non-zero eigenvalues. Alternatively, the rank of A is the difference between its dimension and its total number of zero eigenvalues. From a group theoretic point of view the unitary similarity transformations form a subgroup of the group of similarity transformations. This subgroup is known as the *unitary* group. By the transformation (30) the unitary group is a subgroup of the socalled *orthogonal* group which again is subgroup of the similarity group of transformations. Since rank is invariant under the similarity group there will be no difference between the unitary group and the orthogonal group in this respect. Hence, we shall find it preferable to relate Gilbert's assumption with rank determination under the group of orthogonal transformations.

By Gilbert's assumption the eigenvalues of the quantitative state coefficient matrix A are distinct. Accordingly, at most one of the eigenvalues of A can be zero. It follows, since the dimension of A is X, that the rank of A must be X or X-1. Evidently, the requirement that under the orthogonal group of transformations the rank of A must be at least X-1, is the structural property of Gilbert's assumption to which we desire to give a Boolean representation. Now, as is evident by the conventional rank test (I.5), a column vector B of non-zero entries will suffice even if the rank of A is X-1. However, if A has multiple non-zero eigenvalues in contradistinction to Gilbert's assumption, a single column vector B of non-zero entries will incur a corresponding number of linearly dependent columns in (I.5) with the result that the rank test is not met even if the rank of A is X. Therefore, in the general case, going beyond Gilbert's assumption of distinct eigenvalues, the rank of the A-matrix cannot be considered in isolation without taking into account the effect of the B-matrix (or, in the test for observability, the C-matrix). For the present, however, as an introduction to the more general approach, let us confine ourselves to the structural consequences of Gilbert's assumption.

By Cayley's theorem any finite group of a given order is isomorphic to a group of permutations that is subgroup of the symmetric group of the same order. Invoking this theorem we may give the orthogonal group of transformations a Boolean representation in terms of permutation matrices submitted to the relative product. Thus, if A is the Boolean state coefficient matrix defined by (1) then the quantitative orthogonal representation (31) will now be represented by the qualitative orthogonal representation:

$$(\lozenge P) \vee . \wedge A \vee . \wedge P \qquad (32)$$

where P satisfies (26) by virtue of being a permutation matrix. In a similar manner as the transformation (31) was used to establish the rank of the quantitative matrix A, so we may now use the form (32), in which P has been obtained by quasi-level coding (see chapter 6), to establish the *term rank* of the qualitative matrix A. That is, we use the form (32) to determine the maximal rank that the A-matrix may attain by virtue of its structural pattern of non-zero elements. In other words, we search for the maximal permutation matrix contained in the form (32) as a subset of the non-zero entries since the term rank will be the total number of non-zero elements in this new maximal permutation matrix. Thus, we may conclude, that *the Boolean representation of the structural content of Gilbert's assumption of distinct eigenvalues, is the origin of the term rank test* (see chapter 1).

In the next section we shall enter upon a more detailed investigation of the term rank test aiming at its establishment in the general case going beyond Gilbert's assumption. However, before we do so it will be worthwhile to clarify a fundamental difference in the basic assumptions underlying the reachability test and the term rank test.

The reachability property was established on a nominal scale as an invariant property under the symmetric group of transformations. Hence, the reachability property is invariant not only for all Boolean tensor operations, but also for all Cartesian tensor operations. This is due to the fact that the Cartesian tensors are invariant under the similarity group of transformations which is a subgroup of the symmetric group. Therefore, the reachability property truly describes the invariant structural content of the system under Boolean and Cartesian tensor operations.

The term rank property, on the other hand, is not a tensorial property. First, it was established only for quantitative matrices under the orthogonal group of transformations. In fact, the property cannot be related to Cartesian tensors in general. Secondly, applying Cayley's theorem to give the property a Boolean representation, we had to confine ourselves to the orthogonal subgroup of the symmetric group. Thus, the term rank test is valid only for for a subset of the measurements on the nominal scale. By virtue of being an injection on the Boolean domain it cannot be expressed as an invariant in terms of Boolean tensor operations. Therefore, the term rank is a structural matrix property in a far more narrow sense than the tensorial reachability property.

In the light of Klein's Erlanger Program the difference between reachability and term rank is obvious. Reachability is brought from a nominal scale to a ratio scale. That is, reachability is taken from an invariant scale-form under the symmetric group to an invariant scale-form under the similarity group. Hence, by virtue of the fact that the similarity group is a subgroup of the symmetric group, reachability remains an invariant property. In contradistinction, term rank is moved the opposite way from a subset on the ratio scale to a subset on the nominal scale. On the ratio scale the subset is determined as matrices under the orthogonal group of transformations. The subset on the nominal scale is simply another representation of this orthogonal group of matrix transformations. Thus, we should not expect term rank to appear as a tensorial property.

In the Boolean domain Gilbert's assumption of distinct eigenvalues found its qualitative representation in the structural property of term rank or, as it has recently been called also, *generic rank* (Shields & Pearson, 1976). Basically, determination of the term rank of a Boolean matrix M may take place along either of two avenues. One approach, made popular in connection with the solution of the socalled *assignment problem*, amounts to an identification of the maximal permutation matrix contained in M by virtue of its non-zero entries (Ford & Fulkerson, 1962). The other approach, lately reestablished also from the viewpoint of generic analysis, is tantamount to the establishment of the maximal zero submatrix contained in M by virtue of its zero entries (Ore, 1962); Shields & Pearson, 1976). *)

Apart from their obvious emphasis on the dual properties of the entries of M, the two approaches are distinguished by the fact that they are commonly founded on a theorem proved independently by G. Frobenius and D. König at the early beginning of this century (Churchman, et al., 1957). According to a footnote in König's classical work on graph theory from 1936 it appears that Frobenius, to say the least, was less than enthusiastic about the graph-theoretical approach (König, 1950). In fact, neither the proofs by Frobenius nor by König seem to have escaped criticism judged by the fact that many combinatorial proofs are known of their theorem (Ford & Fulkerson, 1962). It is a common observation that to make a mathematical concept operational in the computational sense the algorithm for its determination may often be devised by consideration of one of its proofs. In this respect König's course of action is particularly amenable because it is based on the geometrically lucid and intuitively simple notion of an *alternating path* of a digraph (Petersen, 1891; Ore, 1962).

Placing Petersen's alternating path method from 1891 at the foundation of the Frobenius-König theorem, permits us to determine the term rank by a purely graph-theoretical procedure which, in its most simple form, amounts to a visual inspection of a digraph. It follows that this approach may be implemented by an APL function in a conceptually straight-forward manner. Having thus based the term rank criterion on the same graph-theoretical foundation as the reachability criterion we see that the main difference is that the former is concerned with alternating paths whereas the latter relates to directed paths. Obviously, since an alternating path is a more narrow concept than a directed path we cannot expect in general that alternating paths remain invariant under the symmetric group of transformations preserving directed paths. In fact, as it turns out, alternating paths remain invariant only under the alternating group which is the subgroup of the symmetric group defined by the set of all *even* permutations. Hence, since only a subset of the measurements on the nominal scale are involved, term rank is not a Boolean tensor property.

5.1 The Alternating Path Method

Let the *independent* entries of a Boolean matrix M of arbitrary shape denote a set of non-zero entries, no two of which lie on the same *line* (i.e., row or column). By the Frobenius-König theorem the maximum number of independent entries of M equals the minimum number of lines (rows or columns) containing all the non-

---

*) Unfortunately, according to a correction kindly communicated to the authors by Drs. Shields & Pearson, their published algorithm implementing this idea was shown later not to work in general. It is thought-provoking that subsequent literature takes this algorithm for granted without a comment simultaneously as it scrutinizes the underlying mathematical proof in order to abbreviate and improve it. Also, it illustrates the need for an operational approach complementary to current tradition.

zero entries of M. The maximum number of independent entries or, equivalently, the minimum number of covering lines, establishes the term rank of M. Evidently, the maximum number of independent entries forms a matrix P of the shape of M which in each row and column has *at most* one entry unity and all others zero. By a generalization of the conventional terminology any such matrix P is called a *permutation matrix* (Ryser, 1963). We may say, therefore, that the term rank is determined by the total number of unit entries in the maximal permutation matrix P contained in M. It is noteworthy that, in contradistinction to the conventional conception, the maximal permutation matrix P will be rectangular if that is the shape of M.

It is a wellknown fact that a Boolean matrix M of arbitrary shape may be represented by a *bipartite* digraph. That is, let the set of row indices and the set of column indices of M represent each of two disjoint sets of vertices such that the unit entries of M depict the total set of edges uniformly directed from the one vertex set to the other. Clearly, a single edge of this digraph may be said to establish a *matching* of the vertex of a row index with that of a column index. It follows that the edges, defined by a maximal permutation matrix P contained in M, assign a *maximal matching* or one-to-one correspondence between a subset of row indices and a subset of column indices. Thus, in addition to the term rank the maximal permutation matrix P will provide a qualitative solution to the socalled *assignment problem* in which a maximal matching is desired between two disjoint sets of entities.

Perhaps, the most wellknown search procedure for determination of a maximal permutation matrix P, is the *assignment algorithm* developed by Ford & Fulkerson. This algorithm which is based on an iterative determination of the minimum number of covering lines, is usually given a purely algebraic formulation in terms of row and column scannings of the given matrix M (Ford & Fulkerson, 1962). The process of labeling rows and columns of M in this algorithm differs basically from the approach adopted by Shields and Pearson. Thus, the latter determine indirectly the minimum number of covering lines by finding the corresponding maximal zero submatrix in a finite sequence of permutation operations (Shields & Perason, 1976). In either case, we end up with an algebraic search process of M that it is difficult to visualize in simple geometrical terms. To reap the benefits of the insight provided by a geometrical interpretation we shall adopt an approach that differs from the above by being based on the graph-theoretical concept of an alternating path. That is, our aim is to explicit introduce the alternating path method as the mathematical foundation of the term rank determination. However, it will be necessary first to establish the necessary graph-theoretical background (Petersen, 1891; König, 1951; Ore, 1962).

The alternating path method will determine a maximal permutation matrix P contained in a Boolean matrix M of arbitrary shape. Yet, to enable us to compare the representation of M by a bipartite digraph with the conventional digraph representation of M as a successor matrix we shall find it convenient for purely illustrative reasons to confine ourselves to square matrices M. In other words, without any loss of generality we utilize by this proviso on M that any conventional digraph has an equivalent representation as a bipartite digraph. This one-to-one correspondence between the two equivalent digraph representations makes it possible to translate results from one representation to the other. In the following we shall take advantage of this fact since it will permit us to establish a maximal permutation matrix P directly from the bipartite digraph by visual inspection and then interprete the result relative to the corresponding successor digraph. It is noteworthy that, whereas it is easy to translate the result to the conventional digraph, it is often error-prone to work directly on the latter.

At the outset, to make the discussion specific, consider a Boolean matrix M, say:

$$\rho\,\square\!\leftarrow\! M$$

```
1 0 1 0 0
1 0 0 0 1
0 0 0 1 0
1 1 0 0 1
0 1 0 0 1
5 5
```

(33)

the conventional and bipartite successor digraphs of which are depicted in Fig. 4. Thus, in the bipartite digraph of Fig. 4B we have, by virtue of conceiving M as a successor digraph, that the unprimed vertex numbers at the top represent column indices. This dependency upon the successor relationship is further emphasized by the designations *OUT* and *IN*. It is readily seen that the edges of the two digraphs are in one-to-one correspondence.

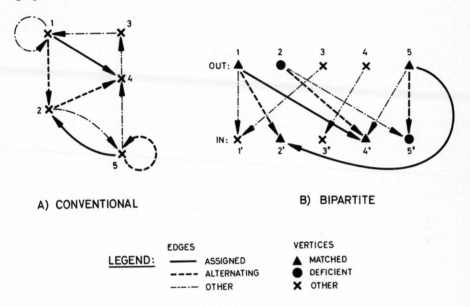

A) CONVENTIONAL   B) BIPARTITE

LEGEND:

EDGES		VERTICES	
——	ASSIGNED	▲	MATCHED
- - - -	ALTERNATING	●	DEFICIENT
—·—·	OTHER	✕	OTHER

Fig. 4. The Concept of an Alternating Path

Let us now assume that we have assigned a partial matching defined by some permutation matrix P contained in M:

$$\rho\,\square\!\leftarrow\! P$$

```
0 0 0 0 0
0 0 0 0 1
0 0 0 0 0
1 0 0 0 0
0 0 0 0 0
5 5
```

(34)

The two non-zero entries of this matrix are shown in Fig. 4 as heavy-lined edges. To answer the question whether or not these two edges make up a maximal matching in the bipartite digraph we shall introduce the concept of an alternating path devised by Julius Petersen (Petersen, 1891; König, 1950; Ore, 1962). It is interesting that, founding the determination of the term rank on this concept, we have come round a full circle since Petersen conceived it as a spin-of from his work on the group-theoretical foundation of algebraic equations (Petersen, 1877).

Essentially, an *alternating path* of a digraph is a topological path of non-repeated edges any consecutive pair of which are oppositely directed. Relative to a matching, specified by a subset of the edges of a bipartite digraph, it is a topological path the edges of which belong alternately to the matching subset and its non-matching complement, usually beginning with an edge belonging to the latter subset. In Fig. 4 these two kinds of edges are called *assigned* respectively *alternating* edges. Let us now distinguish, in a bipartite digraph like Fig. 4B, between vertices that are *matched* and those that are *deficient* or *unmatched*. Clearly, if an alternating path with respect to the assigned edges of a bipartite digraph begins and ends at a pair of deficient vertices, it can be used to derive a new assignment by simply interchanging the rôles of the assigned and the alternating edges. Hence, by this procedure the number of assigned edges will be increased by one. Thus, considering the alternating path which in Fig. 4B begins at the deficient vertex 2 and ends at the deficient vertex 5', it is seen that it consists of three alternating edges (dotted lines) and two assigned edges (heavy lines). Interchanging the rôles of the assigned and the alternating edges of this path the permutation matrix P of (34) is transformed into:

$$
\begin{array}{c}
\rho \square \leftarrow P \\
\begin{array}{ccccc}
0 & 0 & 0 & 0 & 0 \\
1 & 0 & 0 & 0 & 0 \\
0 & 0 & 0 & 0 & 0 \\
0 & 1 & 0 & 0 & 0 \\
0 & 0 & 0 & 0 & 1 \\
5 & 5 & & &
\end{array}
\end{array}
\tag{35}
$$

Clearly, the advantage of this rematching procedure, is that we increase the number of matched vertices by simply adding new vertices to the set of already matched vertices. It is readily appreciated that a similar increased matching can be obtained by simply introducing any edge connecting a pair of deficient vertices. Thus, the edge interconnecting the deficient vertices 3 and 1' in Fig. 4B, illustrates the point in question:

$$
\begin{array}{c}
\rho \square \leftarrow P \\
\begin{array}{ccccc}
0 & 0 & 1 & 0 & 0 \\
1 & 0 & 0 & 0 & 0 \\
0 & 0 & 0 & 0 & 0 \\
0 & 1 & 0 & 0 & 0 \\
0 & 0 & 0 & 0 & 1 \\
5 & 5 & & &
\end{array}
\end{array}
\tag{36}
$$

For this reason we shall consider an edge the vertices of which are deficient to represent a special case of the concept of an alternating path. With a terminological loan from set theory we shall denote any such single-edge alternating path a *singleton*. The fact that a singleton increases the number of matched vertices by two, is of particular importance from a computational point of view.

In general, therefore, the existence of an alternating path opens up for a rematching increasing the number of assigned edges by one edge. It is a well-known fact that *a matching is maximal if and only if the digraph contains no alternative paths relative to the assigned edges* (Ore, 1962). It follows that the problem of finding a maximal matching or permutation matrix is tantamount with the problem of identifying alternating paths whenever they exist.

5.2 Alternating Path Deformations

Obviously, the problem of identifying alternating paths, may be solved by visual inspection of the bipartite digraph representing the Boolean matrix M under consideration. For example, if M is specified by (33) we see from Fig. 4 that a maximal permutation matrix P may be:

$$\begin{array}{cccccc}
0 & 0 & 1 & 0 & 0 \\
0 & 0 & 0 & 0 & 1 \\
0 & 0 & 0 & 1 & 0 \\
1 & 0 & 0 & 0 & 0 \\
0 & 1 & 0 & 0 & 0 \\
\end{array}$$
$$\begin{array}{cc} 5 & 5 \end{array}$$

(37)

The corresponding maximal matching is depicted in the bipartite subgraph of Fig. 5A the assigned edges of which are represented alternatively by the successor subgraph of Fig. 5B. Though, the two digraph representations do not completely agree since in Fig. 5A we have added a subset of alternating edges (indicated by dotted lines) to the set of assigned edges. It is noteworthy that this subset of alternating edges forms a socalled *cyclic* alternating path together with a subset of the assigned edges. That is, an alternating path which begins in an alternating or non-assigned edge and returns to the initial vertex in an assigned edge. Evidently, a cyclic alternating path will contain an even number of edges since its number of alternating edges equals its number of assigned edges. It follows, as originally pointed out by Petersen, that a new maximal matching may be obtained by interchanging the rôles of alternating and assigned edges in any such cyclic path. This operation whereby a

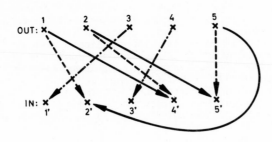

A) MAXIMAL CYCLIC PATH

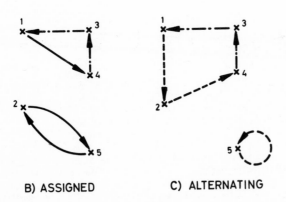

B) ASSIGNED   C) ALTERNATING

Fig. 5. Cyclic Deformation

matching or permutation matrix is transformed into another, is called a *cyclic deformation* with respect to the cyclic alternating path in question (Ore, 1962). Thus, by the cyclic deformation specified by the cyclic path of Fig. 5A the maximal permutation matrix (37) is turned into:

$$
\begin{array}{c}
\rho\square \leftarrow P \\
\begin{array}{ccccc}
0 & 0 & 1 & 0 & 0 \\
1 & 0 & 0 & 0 & 0 \\
0 & 0 & 0 & 1 & 0 \\
0 & 1 & 0 & 0 & 0 \\
0 & 0 & 0 & 0 & 1 \\
\end{array} \\
5 \quad 5
\end{array}
\tag{38}
$$

The successor subgraph representation of this maximal matching is shown in Fig. 5C. Comparison of Fig. 5B and 5C reveals that the conventional successor (or predecessor) digraph representations of the two permutation matrices (37) and (38) exhibit a characteristic patterns of disjoint topological paths. Each such directed path is called a *cycle* since on it any pair of vertices are mutually reachable. It is easy to verify that the disjoint cycles of Figs. 5B and 5C are but geometrical illustrations of the algebraic formulation in cyclic notation of the two permutations (37) and (38) in terms of products of disjoint cycles.

Since any permutation can be given in cyclic notation as a product of disjoint cycles, we would expect that the conventional successor (or predecessor) digraph representation of a permutation matrix will always be a set of disjoint cycles. However, this is true only if we maintain the classical conception of a permutation matrix as a square matrix which in each row and column has a single entry unity and all others zero. In the generalized form considered here the conventional digraph representation of permutation matrices may display an alternative characteristic structure to be discussed shortly. Though, before we enter upon this problem it will be of interest briefly to introduce a few additional concepts related to the structural characterization of conventional digraphs.

Perhaps, the most basic invariancy of a digraph, is whether or not its structure contains cycles. This is due to the fact that the absence or presence of cycles gives rise to a fundamental difference in the reachability properties. Terminologically, this difference is asserted by the distinction between acyclic and cyclic digraphs. Thus, an *acyclic* digraph is characterized by the fact that if vertex I is reachable from vertex J, then vertex J cannot simultaneously be reachable from vertex I. In contradistinction, a *cyclic* digraph will contain at least one pair of vertices I and J that are simultaneously mutually reachable. A special case hereof may be a socalled *loop* which makes a vertex I reachable from itself.

To illustrate the difference between acyclic and cyclic digraphs consider the two Boolean matrices M1 and M2:

$$
\begin{array}{cc}
\begin{array}{c}
\rho\square \leftarrow M1 \\
\begin{array}{cccc}
0 & 0 & 0 & 0 \\
1 & 0 & 0 & 0 \\
1 & 1 & 0 & 0 \\
0 & 0 & 1 & 0 \\
\end{array} \\
4 \quad 4
\end{array}
&
\begin{array}{c}
\rho\square \leftarrow M2 \\
\begin{array}{cccc}
0 & 0 & 1 & 1 \\
1 & 0 & 1 & 1 \\
0 & 1 & 0 & 0 \\
0 & 1 & 0 & 0 \\
\end{array} \\
4 \quad 4
\end{array}
\end{array}
\tag{39a\&b}
$$

Conceived as nodal successor matrices M1 defines the acyclic digraph of Fig. 6A whereas M2 specifies the cyclic digraph of Fig. 6B. At this point the reader may have noticed that *any* pair of vertices in Fig. 6B are mutually reachable. A cyclic digraph exhibiting this special property is said to be *maximally strong* or, simply, *strong*. The term "maximal" refers here to the fact that, usually, the mutual reachability derives from the maximum aggregation of interconnected cycles.

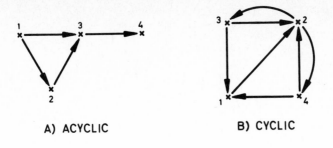

A) ACYCLIC **B) CYCLIC**

Fig. 6. Two Successor Digraphs

A characteristic property of any acyclic digraph, is that it will have at least one *source* (a vertex with no incoming edges) and at least one *sink* (a vertex with no outgoing edges). For example, in Fig. 6A vertex 1 is a source and vertex 4 is a sink. The implication hereof is that the defining Boolean matrix, here M1 of (39a), will contain at least one zero row and one zero column. Evidently, the corresponding maximal permutation matrix will exhibit the same characteristic pattern of zero rows and columns. Another characteristic property of the matrix of an acyclic digraph, is that it will be *nilpotent* by virtue of the fact that the maximal topological distance in the digraph is finite. Consequently, by a corresponding set of row and column permutations it is possible to rearrange the matrix into a *strictly triangular* form with zero entries on and above (or below) the main diagonal. A cyclic digraph may also have sources and sinks. However, if, as illustrated by M2 of (39b), the cyclic digraph is strong, sources and sinks are excluded. On the other hand, the matrix of a cyclic digraph cannot be brought into a strictly triangular form by permutations since clearly it must have elements on or on either side of the main diagonal. Thus, whether the maximal permutation matrix corresponding to a cyclic digraph will have zero rows and columns, must be decided from case to case. To illustrate, a pair of maximal permutation matrices corresponding respectively to the two matrices M1 and M2 of (39) may be the following:

$$\rho\Box\leftarrow P1$$

$$\begin{matrix} 0 & 0 & 0 & 0 \\ 1 & 0 & 0 & 0 \\ 0 & 1 & 0 & 0 \\ 0 & 0 & 1 & 0 \end{matrix}$$

$$\rho\Box\leftarrow P2$$

$$\begin{matrix} 0 & 0 & 0 & 1 \\ 1 & 0 & 0 & 0 \\ 0 & 1 & 0 & 0 \\ 0 & 0 & 0 & 0 \end{matrix} \quad \text{(40a\&b)}$$

The very fact that in either case we have only three independent (non-zero) entries, demonstrates that the maximal matchings, defines by M1 and M2, are completely unaffected by the fundamental differences in the reachability properties of the corresponding digraphs Fig. 6A and B. In general, since maximal matching (and, hence, term rank) is independent of reachability, it is not a Boolean tensor property.

This observation ties up with the remark, made previously, that the digraph representation of the generalized permutation matrices may exhibit a characteristic structure alternative to that of disjoint cycles. To demonstrate the nature of this alternative structure we may consider Fig. 7A which depicts the conventional and the bipartite digraphs describing the alternating paths defined by the generalized permutation matrix P2 of (40b). In the latter bipartite digraph is indicated (by heavy lines) an alternating path interconnecting the

58

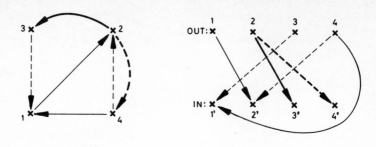

A) EDGE DISJOINT DEFICIENCY PATHS

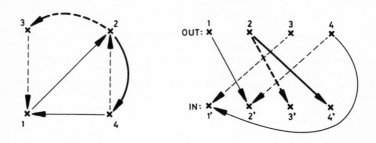

B) TRANSFORMED DEFICIENCY PATHS

Fig. 7. Deficiency Path Deformation

three vertices 3', 2, and 4'. This path consists of an assigned edge (the full line) and an alternating edge (the dotted line). Since it ends at an unmatched or deficient vertex 4' it is called a *deficiency path* for the maximal matching P2. In general it is evident that if a maximal matching is defined in terms of edge disjoint deficiency paths then each such alternating path must be of even length. Accordingly, if the rôles of the assigned and the alternating edges are interchanged in a deficiency path, we arrive at a new maximal matching. This operation which is called a *deficiency path deformation*, is exemplified in Fig. 7B. Obviously, by this operation the maximal permutation matrix (40b) is transformed into:

$$\rho\square{\leftarrow}P2$$

$$\begin{matrix} 0 & 0 & 0 & 1 \\ 1 & 0 & 0 & 0 \\ 0 & 0 & 0 & 0 \\ 0 & 1 & 0 & 0 \end{matrix} \tag{41}$$

$$\text{4 4}$$

Inspection of the bipartite digraphs of Fig. 7A and 7B reveals that in addition to the interchange of rôles of the edges a similar correspondence is established between vertices which, like vertex 3' and vertex 4', alternate in appearance as respectively matching and non-matching or deficient vertices. Clearly, any such pair of corresponding deficient vertices, defined by a maximal permutation and a transformation hereof by a deficiency path deformation, must be the endpoints of an edge disjoint alternating path of even topological length. This, very briefly, is the structural alternative to disjoint cycles in the digraph representation of generalized permutation matrices. In conclusion, therefore, we see that *any maximal matching can be transformed into any other maximal matchings by deformations with respect to deficiency paths and by cyclic deformations* (Ore, 1962 [15]).

The fact, discussed previously, that maximal matching is not a Boolean tensor property, is further demonstrated by considering the reachability properties of the successor digraph representations of Fig. 7. Thus, in Fig. 7A the maximal permutation matrix (40b) appears as an open topological path interconnecting all vertices, whereas in Fig. 7B the transformed maximal permutation matrix (41) forms a cycle interconnecting but a subset of the vertices. Yet, even if the property of maximal matching is not a Boolean tensor property and, hence, is not invariant under the *symmetric* group (or the group of *all* permutations), it should be realized that maximal matching is invariant under a subgroup hereof, namely the *alternating* group (or the group of all *even* permutations). This explains why alternating paths of *even* topological lengths appear as an invariant property of the bipartite digraph representation of maximal matchings.

The distinction between invariant properties under the symmetric group and invariant properties under the alternating group, accounts for the difference between the tensorial properties of measurements on the nominal scale and the structural matrix properties injected from the ratio scale on a subset of the measurements on the nominal scale. However, *symmetry* may disturb this picture by extending the invariancy of the structural matrix properties from permutations under the alternating group to permutations under the symmetric group. This also elucidates the fundamental reason why, in the formulation and solution of matrix equations, the appearance of symmetrical matrices has come to play such a dominant role.

5.3 Assigning a Maximal Permutation Matrix

The alternating path method is based, as we have seen in the previous sections, on an intuitively simple, yet fundamental graph-theoretical conception. Still, in spite of the importance attached to this method by mathematicians like König*) and Ore, the present authors have failed to found in the literature any algorithm directly implementing this method. Of course, we take here exception to algorithms which, like the Ford and Fulkerson assignment algorithm, is founded theoretically, at least indirectly, on the alternating path method (Ford & Fulkerson, 1962). For this reason we shall find it of interest to give an APL implementation of the essential part of the alternating path method in its basic form.

The crucial point in the method is the establishment of a criterion for the existence of at least one alternating path with the property that it can be rematched. That is, that the number of assigned edges can be increased by one. By a terminological loan from the Ford & Fulkerson assignment algorithm the satisfaction of this criterion will be called a *breakthrough*. For purely explanatory reasons the procedure implementing this criterion, will be given a formulation emphasizing the conceptual aspects at the cost of computational efficiency.

Essentially, the problem is to establish an identification of some alternating path from which a new and increased assignment may be derived by the combined tracing and rematching procedure. Clearly, any such alternating path will exhi-

---

*) Thus, in 1936 in his classical work on graph theory König (1950) wrote: *"Diese Abhandlung von Petersen, an der auch Sylvester beteiligt ist, ist sicherlich eine der bedeutendsten Arbeiten über Graphenteorie, scheint aber mehr als 25 Jahre lang fast gänzlich unbeachtet geblieben sind."* (This thesis by Petersen, to which also Sylvester has contributed, is doubtless the most important work in graph theory, though it has been completely ignored for more than 25 years.)

bit two characteristic properties. First, the initial vertex and the final vertex must both be deficient. Secondly, it must consist of an odd number of edges beginning and ending with an alternating edge. That is, it must be either a singleton or some expansion of the pattern of successive edges: an alternating edge of one orientation, an assigned edge of the opposite orientation, and an alternating edge of the first orientation. It is evident that if we reverse the orientations of the assigned edges, say, in such an alternating path, the problem of establishing the condition of a breakthrough is reduced to the wellknown problem of determining whether one deficient vertex is reachable from another deficient vertex.

To give a matrix formulation of how this transformation of the problem may be accomplished consider a Boolean matrix M of arbitrary shape defining a finite bipartite digraph. Further, let us assume that in this digraph we have found a non-empty subset of assigned edged specified by the unit entries of a generalized permutation matrix P of the shape of M. Now, the assigned edges of P may define alternating paths of different topological lengths together with such unit entries of M that specify the appropriate alternating edges. Reversing the orientations of the assigned edges, say, is simply a matter of introducing the transpose of P denoted $\lozenge P$. Hence, to establish a breakthrough we must investigate the reachability property over the following topological paths of increasing lengths:

$$
\begin{aligned}
&M\\
&M\vee.\wedge(\lozenge P)\vee.\wedge M\\
&M\vee.\wedge(\lozenge P)\vee.\wedge M\vee.\wedge(\lozenge P)\vee.\wedge M\\
&\cdots\cdots\cdots\\
&M\vee.\wedge(\lozenge P)\vee.\wedge M\vee.\wedge\cdots\cdots\vee.\wedge(\lozenge P)\vee.\wedge M
\end{aligned}
\tag{42a}
$$

By virtue of the fact that the relative product is associative (42a) may be rewritten:

$$
\begin{aligned}
&M\vee.\wedge I\\
&M\vee.\wedge((\lozenge P)\vee.\wedge M)\\
&M\vee.\wedge((\lozenge P)\vee.\wedge M)\vee.\wedge((\lozenge P)\vee.\wedge M)\\
&\cdots\cdots\cdots\\
&M\vee.\wedge((\lozenge P)\vee.\wedge M)\vee.\wedge\cdots\cdots\vee.\wedge((\lozenge P)\vee.\wedge M)
\end{aligned}
\tag{42b}
$$

where I is a unit matrix the dimension of which is defined by the dimension of the square Boolean matrix $((\lozenge P)\vee.\wedge M)$.

By comparison with (I.54) we see that if we neglect the initial factor M in each line of (42b) the remainder expressions are simply the powers: 0, 1, 2, etc. of the square matrix $((\lozenge P)\vee.\wedge M)$. Thus, applying (I.55) an arbitrary line of (42b) may be expressed:

$$
MJ \leftarrow M\vee.\wedge((\lozenge P)\vee.\wedge M) \ \underline{OR\triangle AND} \ J
\tag{42c}
$$

where MJ takes the dimensions of M. The importance of matrix MJ, is that if it contains a unit entry assigning a deficient row index to a deficient column index, then an alternating path exists of a topological length $(1+2\times J)$ such that the number of assigned edges can be increased by one. However, since we are interested only in the existence of this condition along an alternating path of arbitrary length we may as well consider the union of all alternating paths of different lengths. That is, combining the lines of (42b) by disjunction, moving the factor M outside the parentheses, and invoking the function $\underline{REACHABILITY}$ defined in section I.4.2 we arrive at the expression:

$$
RM \leftarrow M\vee.\wedge \ \underline{REACHABILITY} \ (\lozenge P)\vee.\wedge M
\tag{42d}
$$

specifying the generalized reachability matrix RM on which we have founded the criterion of a breakthrough. This criterion, to be sure, works in exactly the same manner as that described above for matrix MJ. To illustrate, consider the application of (42d) to matrix M of (33) and P of (34) the result of which should be compared with Fig. 4:

```
        ρ□←M                          ρ□←P
   1 0 1 0 0                     0 0 0 0 0
   1 0 0 0 1                     0 0 0 0 1
   0 0 0 1 0                     0 0 0 0 0
   1 1 0 0 1                     1 0 0 0 0
   0 1 0 0 1                     0 0 0 0 0
   5 5                          5 5
```

```
     ρ□←(◊P)∨.∧M                 ρ□←REACHABILITY (◊P)∨.∧M
  1 1 0 0 1                   1 1 0 0 1
  0 0 0 0 0                   0 1 0 0 0
  0 0 0 0 0                   0 0 1 0 0
  0 0 0 0 0                   0 0 0 1 0
  1 0 0 0 1                   1 1 0 0 1
  5 5                        5 5
```

(43a-e)

```
       ρ□←M∨.∧REACHABILITY (◊P)∨.∧M
   1 1 1 0 1
   1 1 0 0 1
   0 0 0 1 0
   1 1 0 0 1
   1 1 0 0 1
   5 5
```

Clearly, as revealed by visual inspection, we have a breakthrough.

This procedure is implemented in the function *BREAKTHROUGH*. Its arguments are the quantitative matrix M and an assumed conforming permutation matrix P, identifying by its unit entries a set of assigned edges in the bipartite digraph defined by M. The result L of the function is a Boolean vector of dimension 1↑ρM. The unit entries of L designate the row indices, and hence the vertices, for which we have a breakthrough. Thus, if all entries of L are zero then P is a maximal permutation matrix.

```
           ∇ BREAKTHROUGH [□] ∇

        ∇ L←P BREAKTHROUGH M;RM
  [1]   ⋒
  [2]   ⋒      AN ASSUMED PERMUTATION MATRIX "P",
  [3]   ⋒      SPECIFIED FOR A QUANTITATIVE MATRIX
  [4]   ⋒      "M", IS TESTED FOR BREAKTHROUGH.
  [5]   ⋒
  [6]   M←M≠0
  [7]   RM←M∨.∧REACHABILITY(◊P)∨.∧M
  [8]   L←(∨≠P)<(~∨/P)∨.∧RM
        ∇
```

Invoking the function with the arguments specified by the example in (43a-e):

$$\rho\square\leftarrow P \; \underline{\mathit{BREAKTHROUGH}} \; M$$

0 1 1 1 0   (44)

5

indicates a breakthrough for row indices 2, 3, and 4.

The essence of the procedure implemented in $\underline{\mathit{BREAKTHROUGH}}$, is that it establishes the criterion for increasing the term rank of P. However, it does not specify a breakthrough in the sense that an actual such alternating path is identified by a tracing procedure and a corresponding rematching of the edges in this path is performed. Evidently, to establish a maximal permutation matrix both of these steps must be performed iteratively. Thus, repeatedly working through these two steps the number of assigned edges, defined by the unit entries of matrix P, is increased iteratively by determining and tracing any would-be alternating path that may be produced from each of the remaining deficient vertices. $\underline{\mathit{ASSIGN}}$ is a heuristic APL function that implements this two-step approach.

```
          ∇ ASSIGN [□] ∇

        ∇ P←ASSIGN M;X;ΔX;Y;AUX
[1]    ⍝
[2]    ⍝    ASSIGNMENT OF A MAXIMAL PERMUTATION
[3]    ⍝    MATRIX "P" TO A GIVEN MATRIX "M".
[4]    ⍝
[5]      M←M≠0
[6]    INITIAL:P←(ρM)ρ0
[7]    ΔRANK:X←~∨/P
[8]      AUX←(¯1↑ρM)ρ0
[9]    SEARCH:Y←X∨.∧M
[10]   BREAKTHROUGH:→(~∨/AUX<Y)/0
[11]    AUX←<\(∨≠P)<(AUX<Y)
[12]    →(∨/AUX)/TRACEPATH
[13]    X←X∨P∨.∧Y
[14]    AUX←Y
[15]    →SEARCH
[16]   TRACEPATH:ΔX←<\X∧M∨.∧AUX
[17]    X←ΔX<X
[18]    Y←(⍉P)∨.∧ΔX
[19]    P←(ΔX∘.∧Y)<P∨(ΔX∘.∧AUX)
[20]    AUX←Y
[21]    →(∨/AUX)/TRACEPATH
[22]    →ΔRANK
        ∇
```

The argument M of this function, which is origin independent, should be a rectangular or square quantitative or Boolean matrix. The result P will be a maximal permutation matrix of the same dimensions as M. It is wellknown that in many APL applications it is found expedient to specify, say, a single-element matrix as a scalar or a row-matrix as a vector. Conceptually this disagrees with our fundamental notion of a permutation matrix as the assignment of row indices to column indices (or vice versa) of a Boolean matrix. Accordingly, in such cases it is required first to reshape the argument into the proper matrix rank. A few illustrations on how to use the function are given in (45).

```
        ρ□←M                        ρ□←ASSIGN M
   0  1  1  0                  0  1  0  0
   1  1  1  1                  1  0  0  0
   0  0  1  0                  0  0  1  0
   0  0  1  0                  0  0  0  0
   4  4                        4  4


        ρ□←N                        ρ□←ASSIGN N
   1  1  1  1  0  1            1  0  0  0  0  0
   1  0  0  1  1  0            0  0  0  1  0  0
   1  1  1  1  1  1            0  1  0  0  0  0
   0  1  1  0  0  1            0  0  1  0  0  0
   4  6                        4  6                    (45a-e)


        ρ□←⍉N                       ρ□←ASSIGN ⍉N
   1  1  1  0                  1  0  0  0
   1  0  1  1                  0  0  1  0
   1  0  1  1                  0  0  0  1
   1  1  1  0                  0  1  0  0
   0  1  1  0                  0  0  0  0
   1  0  1  1                  0  0  0  0
   6  4                        6  4


        ρ□←S                        ρ□←ASSIGN 1 1ρS
   1                          1
                              1  1


        ρ□←V                        ρ□←ASSIGN (1,ρV)ρV
   0  0  1  1                  0  0  1  0
   4                          1  4
```

At this point a few, more technical comments may be appropriate for the benefit of readers who, based on previous experience with APL, desire to study the function in more details. These comments, we emphasize, are concerned only with the interpretation of some basic APL operations of principal interest.

The two step iterative procedure implemented by *ASSIGN* naturally subdivides this function into two parts. The first part, statements [6]-[15], establishes the breakthrough criterion while the second part, statements [16]-[22], identifies an actual alternating path increasing the term rank by one. The breakthrough criterion, originally specified by the function *BREAKTHROUGH*, has now been given an alternative formulation such that the breakthrough of the shortest alternating path is identified. The APL operation of decisive importance in this connection is the composite operator <\... introduced in statement [11].

Applied to conforming Boolean arguments the dyadic or binary operation ...<... appears as a Boolean non-implication. Essentially, it removes unit entries of the right arguments which coincide with unit entries of the left argument. It may be used, therefore, to update changes in some Boolean argument. Combining this non-implication with the socalled scan operator, results in the composite operator <\... which eliminates all unit entries but the first along the last dimension of its right and only Boolean argument. Thus, if the argument is a Boolean vector this operation identifies the leading entry of the vector. In the form <⌿... (not used in this function) the leading unit entries are determined along the first dimension of the Boolean argument.

The auxiliary variable AUX in statement [11] is a vector the unit entries of which indicate deficient vertices giving rise to a breakthrough. By the composite operator <\... one of these vertices are selected. Tracing backwards from this vertex a specific alternating path is identified among those causing a breakthrough. Simultaneously, the corresponding rematching is performed in statement [19] interchangeably deleting and inserting unit entries in matrix P. In fact, the second part of the function, statements [16]-[22], is but a splitting of the expression (42d) into its component parts so as to permit an iterative backward search of an alternating path.

From a computational point of view the critical part of the function *ASSIGN* are the statements [16]-[22]. Thus, whereas the first part, statements [6]-[15], is polynomial with respect to the number of iterations, the second and critical part is exponential. Further, it cannot be guaranteed that the second part will produce from a breakthrough an alternating path increasing the term rank. Yet, in actual practice no such failure has been encountered.

The computational inefficiency of the function *ASSIGN* makes its application unacceptable in practice except for small-scale systems in the order of less than fifty state-variables. Hence, to overcome this drawback we give an alternative, rather efficient function called *ASSIGN*1. Essentially, this function implements the wellknown principle underlying the Ford & Fulkerson assignment algorithm. Comparison with the previous function *ASSIGN* will reveal that to speed up *ASSIGN*1 we have substituted the use of Boolean search vectors by a pointer technique implemented by means of index vectors.

```
         ∇ ASSIGN1 [□] ∇

         ∇ P←ASSIGN1 M;C;R;Y;X;ΔX;ΔY;BEG;END
    [1]    ⍝
    [2]    ⍝       ASSIGNMENT OF A MAXIMAL PERMUTATION
    [3]    ⍝       MATRIX "P" TO A MATRIX "M" BY THE
    [4]    ⍝       FORD AND FULKERSON METHOD.
    [5]    ⍝
    [6]    M←M≠0
    [7]    INITIAL:P←(ρM)ρ0
    [8]    R←ι1↑ρM
    [9]    C←ι¯1↑ρM
   [10]    ΔRANK:→(∧/∨/P)/0
   [11]    FORWARD:ΔX←BEG←~∨/P
   [12]    Y←(¯1↑ρM)ρ0
   [13]    X←(1↑ρM)ρ0
   [14]    CLABEL:X←X+ΔX
   [15]    ΔY←(0=Y)×(R×ΔX≠0)⌈.×M
   [16]    END←ΔYx~∨≠P
   [17]    BREAKTHROUGH:→(∨/0≠END)/BACKWARD
   [18]    →(∧/0=ΔY)/0
   [19]    RLABEL:Y←Y+ΔY
   [20]    ΔX←P⌈.×C×ΔY≠0
   [21]    →CLABEL
   [22]    BACKWARD:I←⌈/END
   [23]    J←ENDιI
   [24]    TRACEPATH:P[I;J]←1
   [25]    →(BEG[I])/ΔRANK
   [26]    J←X[I]
   [27]    P[I;J]←0
   [28]    I←Y[J]
   [29]    →TRACEPATH
         ∇
```

The outcome of invoking the APL function *ASSIGN*, implementing the alternating path method, is, as we have seen, a generalized permutation matrix P defining a maximal set of independent entries of a Boolean matrix M of arbitrary size. It follows that to find the term rank of M we must add all the unit entries of P. This is done by the double summation:

$$+/+/P \qquad\qquad (46)$$

which yields a non-negative and integral scalar value.

In general, therefore, the *term rank*, denoted T, of a Boolean matrix M of arbitrary shape is determined by the expression:

$$T \leftarrow +/+/\underline{ASSIGN}\ M \qquad\qquad (47)$$

Previously, in section 3.2 we clarified the difference between acyclic and cyclic digraphs in relation to the illustrations of such digraphs given in Fig. 6. It will be of interest here to relate these structural properties to that of term rank. For this purpose let us assume that M is a *square* Boolean matrix. Now, if M represents an *acyclic* digraph it is evident, as illustrated by (40a), that its term rank T must be less than its dimension:

$$T < 1 \uparrow \rho M \qquad\qquad (48)$$

by virtue of the fact that it contains at least a source and a sink. On the other hand, if M represents a *strong cyclic* digraph, the absence of sources and sinks, as demonstrated by (40b), does not guarantee a term rank equal to the dimension of M. In fact, all we can say is that the term rank T will be less than or equal to its dimension:

$$T \leq 1 \uparrow \rho M \qquad\qquad (49)$$

Hence, to decide upon either of these two possibilities we must submit M to a term rank test.

The difference between (48) and (49) will be of importance for the discussion later in part III of the determination of term rank in combination with topological decomposition. There, one of the problems will be to decide upon whether or not the term rank of a square Boolean matrix M will equal its dimension. Clearly, the answer to this problem may be derived by consideration of the corresponding digraph. That is, the answer will be negative if the digraph is acyclic or if it is cyclic containing at least one source or one sink. On the other hand, no definite answer can be given for a strong cyclic digraph. In this particular case, to be sure, the answer must be provided by an explicit determination of the term rank.

5.4 Testing the Augmented System

Considering the state equation (I.1), relating the $\underline{X}$ state variables with their first order time-derivatives and the $\underline{U}$ input or control variables, Lin (1974) established a term rank test for potential state controllability assuming a single input only. Subsequently, this result was generalized to admit multiple inputs (Shields & Pearson, 1976; Glover & Silverman, 1976). In its generalized form the test is concerned with the term rank of the compounded matrix (A,B) produced by catenation of the two coefficient matrices A and B. Thus, to satisfy the test it is required that:

$$\underline{X} = +/+/\underline{ASSIGN}\ A,B \qquad\qquad (50)$$

Though, to admit a combination of this result with the topological decomposition to be undertaken in part III, it will be necessary to redefine the formulation of the compounded matrix argument. In (50) this argument is a rectangular matrix of dimensions:

$$(\underline{X},\underline{X}+\underline{U}) \;=\; \rho A,B \tag{51}$$

Clearly, to permit a representation by a digraph this matrix must be made square by catenation of $\underline{U}$ zero rows. This operation which of course leaves the term rank invariant, may appear like an insignificant algebraic trick. But in fact, it does imply a significant augmentation of the control system having important ramifications into the tensorial formulation of part III. Here, to lay the foundation of this discussion, it will be necessary to adopt an argumentation alternative to that which, in the literature mentioned above, originally led to (50).

Previously, we have seen one example of the fact that, to establish isomorphisms on a nominal scale by injection from measurements on a ratio scale, all the defining transformations or operations must take place on the ratio scale. Thus, it was by this approach that we established the fundamental isomorphism between Gilbert's quantitative criteria and the reachability criteria. Of course, the underlying reason is that measurements on a ratio scale are invariant under the similarity group which is a subgroup of the symmetric group under which measurements on a nominal scale remain invariant. Hence, it is only by injection from the ratio scale into the nominal scale that we can avoid endowing the analogous measurements on the nominal scale with properties beyond such which are defined by the corresponding measurements on the ratio scale.

The formulation of a term rank test as a structural bound on the conventional rank test for state controllability (I.5), falls within this category of correspondences that must be established by injection from the ratio scale. In other words, the argumentation must be based on a purely quantitative derivation of which only the results are analogized on the Boolean domain. It is to the problem of deducing this kind of approach that we now turn.

Essentially, our approach will be founded on the wellknown theorem from conventional matrix algebra that the rank of the product of two conforming matrices M and N, is less than or equal to the minimum rank of the two matrices (Timothy & Bona, 1968):

$$(\underline{RANK} \; M+.\times N) \;\leq\; (\underline{RANK} \; M) \; \mathsf{L} \; \underline{RANK} \; N \tag{52}$$

In this expression the designation $\underline{RANK}$ denotes some APL function that produces, as an explicit result, the rank of its matrix argument. Now, introducing, by (I.8) or (I.9), a non-zero J'th power of a square matrix N, called NJ, it is obvious that:

$$(\underline{RANK} \; NJ) \;\leq\; \underline{RANK} \; N \tag{53}$$

This is the form in which the theorem will be applied in the following.

In passing, permitting a digression, it should be noted that (53) is an invariant under the similarity group defining the elements of N to be measured on a ratio scale. Though, the Boolean analogy of (53), substituting conventional rank by term rank, is *not* an invariant under the symmetric group defining the elements of N to be measured on a nominal scale. Definitely, as we have already seen, term rank in general is invariant only under the alternating group. But what is the meaning of this limitation on the Boolean analogy on (53) in relation to other structural properties that remain invariant under the full symmetric group. To provide some insight into this problem and also to indicate the importance of a separation of these two kinds of structural properties, assume, as an illustrative example, that matrix N represents a strong cyclic digraph. Then, for some finite power J of N, we may very well find that any pair of vertices of the corresponding digraph will be mutually reachable. In other words, NJ will be a full unit matrix, implying an increase of the term rank in defiance of the Boolean analogy of (53). To be sure, by virtue of the fact that the similarity group is a subgroup of the symmetric group, the analogy of (53) will be valid only for Boolean matrices the elements of which form the proper subset of measurements on the nominal scale. It is not difficult to see that in this subset, specifying the injection of (53) into the Boolean domain, we find all Boolean matrices representing acyclic digraphs since any such matrix is nilpotent. Thus, by this digression we have exemplified the limitations of determining structural properties by injection from the ratio scale into the nominal scale. Clearly, this must be seen in contradistinction to the general validity of the process of defining structural properties like reachability on the nominal scale followed by a projection of these properties on the ratio scale (Franksen, 1975).

But let us turn back to our main line of argument. Here, at the outset, let us rewrite the state equation (I.1) so that in conventional mathematical notation it becomes:

$$
\begin{bmatrix} \dot{X} \\ \dot{U} \end{bmatrix} = \begin{bmatrix} A & B \\ 0 & 0 \end{bmatrix} \times \begin{bmatrix} X \\ U \end{bmatrix}
\tag{54}
$$

which implies that the time-derivative of U does not enter the equation since:

$$
\dot{U} \neq f(X,U)
\tag{55}
$$

With the system in this augmented form the coefficient matrix of (54), say F, is usually referred to as the *augmented-system* matrix (Timothy & Bona, 1968):

$$
F = \begin{bmatrix} A & B \\ 0 & 0 \end{bmatrix}
\tag{56}
$$

In APL this matrix may be determined from coefficient matrices A and B by the statement:

$$
F \leftarrow ((\rho A) + {}^{-}1 \uparrow \rho B) \uparrow A , B
\tag{57}
$$

assuming a vector B reshaped previously into the form of a column matrix.

Evidently, F is a square matrix of dimensions:

$$
((\underline{X} + \underline{U}), \underline{X} + \underline{U}) = \rho F
\tag{58}
$$

It follows by (53) that for a non-zero J'th power of F, say FJ, we have:

$$
(\underline{RANK} \ FJ) \leq \underline{RANK} \ F
\tag{59}
$$

where obviously:

$$
(\underline{RANK} \ F) = \underline{RANK} \ A , B
\tag{60}
$$

by virtue of the $\underline{U}$ rows of zeros.

However, the really interesting property of F lies in the non-zero submatrices of its different powers. Thus, in conventional notation we find:

$$\begin{bmatrix} A & B \\ 0 & 0 \end{bmatrix} ; \begin{bmatrix} A^2 & A{\times}B \\ 0 & 0 \end{bmatrix} ; \begin{bmatrix} A^3 & A^2{\times}B \\ 0 & 0 \end{bmatrix} ; \dots ; \begin{bmatrix} A^{P+1} & A^P{\times}B \\ 0 & 0 \end{bmatrix} \tag{61}$$

which immediately reveals, by comparison with (I.4) and (I.5), that the powers of F simultaneously produce the power series of A together with the matrix products involved in the conventional rank test for state controllability. A catenation of all the matrices of (61), therefore, will produce a matrix, here denoted G:

$$G = \begin{bmatrix} A & B & A^2 & A{\times}B & A^3 & A^2{\times}B & \cdots & A^{P+1} & A^P{\times}B \\ 0 & 0 & 0 & 0 & 0 & 0 & \cdots & 0 & 0 \end{bmatrix} \tag{62}$$

which contains as a submatrix among its columns the matrix K (I.5) underlying the conventional rank test.

By (59) the maximum rank of any of the matrix powers of (61) will be the rank of F. To indicate why this must also be the maximum rank of G assume that the rank of F is less than X, say X-1. It follows that we must add a zero eigenvalue to those already determined by the U zero rows of F. Applying Schur's theorem we may bring F into a triangular form with the eigenvalues on the diagonal such that the additional zero eigenvalue appears in the row immediately above the bottom U zero rows. For the square and higher powers of F the row of the additional zero eigenvalue will be identically a zero row. Thence, none of the columns of G deriving from powers of F can increase the rank of G beyond X-1. We may conclude, therefore, that the rank of the composite matrix G established by catenation of matrix powers of F, will equal the rank of F:

$$(\underline{RANK}\ G) = \underline{RANK}\ F \tag{63}$$

Now, assuming that:

$$\underline{X} > \underline{RANK}\ F \tag{64}$$

it follows by (63) and (62) that the rank of the submatrix K, defined by (I.5), will be:

$$\underline{X} > \underline{RANK}\ K \tag{65}$$

implying that the system is *not* state controllable. In other words, the condition (64) establishes a *lower bound* on the conventional rank test for state controllability.

Injection of this result into the Boolean domain requires that we establish a Boolean matrix F representing (56). With the Boolean coefficient matrices A and B specified respectively by (I.49) and (I.58), F may be produced by (57) assuming B given on matrix form. On this basis, by analogy to (64), we arrive at the *term rank criterion for potential state controllability*, namely, that the system must satisfy the condition:

$$\underline{X} = +/+/\underline{ASSIGN}\ F \tag{66a}$$

This, by (I.10), may also be expressed as:

$$(1\uparrow\rho A) = +/+/\underline{ASSIGN}\ F \tag{66b}$$

By the definition (56) it is obvious that (66a) agrees with the original statement (50). Also, it is immediately seen that, from a purely computational point of view, (50) is preferable to (66a) if the term rank is determined for F in toto. However, if we desire to decompose F topologically in order to simplify the determination of its term rank, then, as we shall see in part III, F must be specified on the form (56).

A point of theoretical interest in connection with (66), may be to relate this condition to Gilbert's assumption on distinct eigenvalues. Here, we find by the derivation leading to (64) in comparison with the observation, made previously, that Gilbert assumed the rank of A to be at least X-1, that the term rank test appears as a generalization of his assumption. Indeed, what we have done, has been to substitute his assumption of a specific property of the rank by the assumption of a lower bound on his property.

To illustrate the use of the term rank test for potential state controllability consider the Boolean matrices A and B defined by (1) or (I.56) respectively by (I.66). Applying (57) we find the augmented-system matrix F:

$$\rho\square\leftarrow F\leftarrow((\rho A)+{}^{-}1\uparrow\rho B)\uparrow A,B$$

$$
\begin{array}{cccccc}
0 & 1 & 0 & 0 & 1 & 0 \\
0 & 1 & 0 & 0 & 0 & 1 \\
0 & 0 & 0 & 1 & 0 & 0 \\
1 & 0 & 1 & 0 & 0 & 0 \\
0 & 0 & 0 & 0 & 0 & 0 \\
0 & 0 & 0 & 0 & 0 & 0 \\
6 & 6 & & & &
\end{array}
\tag{67}
$$

which, in accordance with the assumption made in part I, satisfies the condition (66):

$$+/+/\square\leftarrow\underline{ASSIGN}\ F$$

$$
\begin{array}{cccccc}
0 & 1 & 0 & 0 & 0 & 0 \\
0 & 0 & 0 & 0 & 0 & 1 \\
0 & 0 & 0 & 1 & 0 & 0 \\
1 & 0 & 0 & 0 & 0 & 0 \\
0 & 0 & 0 & 0 & 0 & 0 \\
0 & 0 & 0 & 0 & 0 & 0 \\
4 & & & & &
\end{array}
\tag{68a\&b}
$$

$$1\uparrow\rho A$$

$$4$$

The successor digraph depicting the augmented-system matrix F of (67), is shown in Fig. 8. By comparison with Fig. 1 we see that vertices 5 and 6, appearing as sources (no incoming edges), have been added to the system digraph to explicitly represent the input or control variables. In other words, the digraph of state variables in Fig. 1 have been augmented by the dotted edges specifying vertices 1 and 2 as immediate successors of vertices 5 and 6 respectively. It is immediately observed, in general for any augmented-system digraph, that *control variables are represented graph-theoretically by sources appearing as immediate predecessors of a subset of the state-variables*. Of course, the former of these two attributes derives from the assumption (55), whereas the latter stems from the definition of the B-matrix.

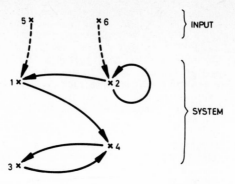

Fig. 8. A Successor Digraph of Control and State-Variables

Turning to the dual problem of formulating a term rank condition for potential observability, it has been stated that by duality to (50) this criterion must be (Davison, 1977).

$$\underline{X} = +/+/\underline{ASSIGN}\ C,[1]A \tag{69}$$

assuming 1-origin. The argument of this expression is produced catenating the A-matrix beneath the C-matrix. The outcome of this catenation is a rectangular matrix of dimensions:

$$((\uparrow\underline{Y}+\underline{X}),\underline{X}) = \rho C,[1]A \tag{70}$$

Evidently, to make this matrix square we must augment it by $\underline{Y}$ columns of zero.

Consideration of the state and output equations (I.1) and (I.2), indicates that in conventional mathematical notation the dual form of (54) must be:

$$\begin{bmatrix} Y \\ \cdot \\ X \end{bmatrix} = \begin{bmatrix} 0 & C \\ 0 & A \end{bmatrix} \times \begin{bmatrix} \dot{Y} \\ X \end{bmatrix} \tag{71}$$

since the time-derivative of the output vector Y does not enter the defining equations. With the system in this augmented form the coefficient matrix of (71), say H, will be the dual of (56):

$$H = \begin{bmatrix} 0 & C \\ 0 & A \end{bmatrix} \tag{72}$$

The basic significance of the *zero-row* form of the coefficient matrix F of (56), is well established in the control theory literature. On the other hand, the dual *zero-column* form of the coefficient matrix M of (72) seem to have attracted less attention. Yet, in classical mechanics the fundamental importance of the zero-column form for the integration of the Lagrangian equations, was recognized as early as in 1877 by E.J. Routh and, somewhat later, in 1884 by H.V.

Helmholtz (Lanczos, 1949). The corresponding variables, like the time-derivative $\dot{Y}$ in (71), were called *absent coordinates* by Routh and *cyclic variables* by Helmholtz. The general accepted terminology of today, is *ignorable variables* which designation was introduced by E.T. Whittaker (1960) in his famous text book from 1904.

Determining matrix powers of H we find corresponding to (61) that these matrices:

$$\begin{bmatrix} 0 & C \\ 0 & A \end{bmatrix} ; \begin{bmatrix} 0 & C{\times}A \\ 0 & A^2 \end{bmatrix} ; \begin{bmatrix} 0 & C{\times}A^2 \\ 0 & A^3 \end{bmatrix} ; \ldots ; \begin{bmatrix} 0 & C{\times}A^P \\ 0 & A^{P+1} \end{bmatrix} \tag{73}$$

contain as non-zero submatrices the terms of the power series of A (I.4) together with the matrix products involved in the conventional rank test for observability (I.6). It follows that catenation of the powers of H beneath each other will produce a matrix corresponding to matrix G of (62). This matrix, evidently, contains as a submatrix among its rows the matrix M (I.6) underlying the conventional rank test. By analogy to the derivation for potential state controllability, therefore, we find that the *term rank criterion for potential observability*, dual to (66), is:

$$\underline{X} = +/+/\underline{ASSIGN}\ H \tag{74a}$$

or, alternatively:

$$(1{\uparrow}\rho A) = +/+/\underline{ASSIGN}\ H \tag{74b}$$

In APL matrix H of (72) may be determined, corresponding to (57) and invoking 1-origin, by the statement:

$$H{\leftarrow}(-(1{\uparrow}\rho C)+\rho A){\uparrow}C,[1]A \tag{75}$$

assuming a vector C reshaped previously into the form of a row matrix. Clearly, by this specification H becomes a square matrix of dimensions:

$$((\underline{Y}+\underline{X}),\underline{Y}+\underline{X}) = \rho H \tag{76}$$

To exemplify an application of the term rank test for potential observability consider the Boolean matrices A and C defined by (1) or (I.56) respectively by (I.83). By (75) we find the augmented matrix H:

```
    ρ□←H←(-(1↑ρC)+ρA)↑C,[1]A
 0  0  1  0  1  0
 0  0  1  0  0  0
 0  0  0  1  0  0
 0  0  0  1  0  0
 0  0  0  0  0  1
 0  0  1  0  1  0
    6  6
```
(77)

which in accordance with the assumption made in part I, satisfies the condition (74):

$+/+/\square \leftarrow \underline{ASSIGN} \ H$

```
0  0  1  0  0  0
0  0  0  0  0  0
0  0  0  1  0  0
0  0  0  0  0  0
0  0  0  0  0  1
0  0  0  0  1  0
4
        1↑ρA
4
```

(78a&b)

The successor digraph representing the augmented matrix H of (77), is given in Fig. 9. Here, to facilitate a comparison with Fig. 1, we have depicted the output variables by vertices 5 and 6 despite the fact that in H they appear as

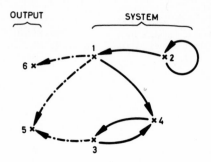

Fig. 9. A Successor Digraph of Output and State-Variables

the first two rows (columns). It follows that the state-variables, specified by the last four rows (columns) of H, are portrayed by vertices labelled exactly as in Fig. 1. Considering vertices 5 and 6, describing the output variables, it is seen that they are sinks appearing as the immediate successors of vertices 1 and 3. By duality to Fig. 8 it is observed in general that *output variables are represented graph-theoretically by sinks appearing as immediate successors of a subset of the state-variables.*

It should be noted that if we desire to use Fig. 9 to determine the reachability property defining potential observability, then, corresponding to the establishment of Fig. 2, the directions of all edges should be reversed. By (2) the matrix defining the converse digraph thus produced, is the transpose $\lozenge H$ of the augmented matrix H.

The procedures defining the term rank tests for potential state controllability and potential observability, are summarized in Table 2. The arguments involved are the Boolean coefficient matrices A, B, and C. The assumption that arguments B and C have matrix form, stated in the first row, may be avoided by a different specification of the augmented matrices F and H. For example, whenever B or C is a vector we may substitute the term invoking ρB or ρC by a 1. The catenation in the specification of the H-matrix presupposes 1-origin.

STATE CONTROLLABILITY	OBSERVABILITY
$2=\rho\rho B$	$2=\rho\rho C$
$F \leftarrow ((\rho A)+{}^{-}1\uparrow\rho B)\uparrow A,B$ $((\underline{X}+\underline{U}),\underline{X}+\underline{U})=\rho F$	$H \leftarrow (-(1\uparrow\rho C)+\rho A)\uparrow C,[1]A$ $((\underline{Y}+\underline{X}),\underline{Y}+\underline{X})=\rho H$
$\underline{X}\leftarrow 1\uparrow\rho A$	
$P \leftarrow \underline{ASSIGN}\ F$	$P \leftarrow \underline{ASSIGN}\ H$
$\underline{X}\ =\ +/+/P$	

Table 2. The Term Rank Tests

CONCLUSION

In this part, emphasizing the striking analogy between Gilbert's formulation and its graph-theoretical generalization in terms of digraph reachability and alternating paths, the qualitative concepts of potential controllability and observability were established as the evident structural counterparts of the quantitative concepts of state controllability and observability. Determination of whether or not a system was potentially controllable or observable, required the system to satisfy simultaneously two independent conditions called the reachability test and the term rank test. These two conditions were explained in terms of digraph properties which in simple cases can be established by visual inspection. Alternatively, regarding more complicated situations, the digraph properties, making up the two criteria, were formulated in APL notation the use of which were demonstrated by numerical examples executed interactively on an APL-terminal.

The characterization of controllability and observability by qualitative design rules is our general aim in this presentation. In an intuitively obvious way this viewpoint appears to complement the conventional mode whereby these concepts are described in qualitative terms. We may agree that matrix rank is a quantitative property while reachability and alternating paths (or term rank) are qualitative properties. Yet, feeling uneasy about the fact that both quantitative and qualitative properties can be determined by calculation, we may prefer to designate the latter structural rather than qualitative. Still, even if digraphs manifest themselves as the skeletons of scientific concepts, it is disturbing that only some, but not all of the digraph properties may be represented by Boolean tensors. In fact, why is reachability more structural than alternating paths or term rank. Indeed, to grade the nature of qualitativeness stretches the imagination beyond common sense, and above all, this grading is required to reflect the enduring core concealed behind the quantitative facade that originally was adopted in the definitions of the concepts.

Repeatedly in our previous presentation we have endeavoured to show that the general answer to this problem, is to invoke the Erlanger Program in terms of the invariant scale-forms of measurements. To wit, a physically significant and consistent classification over the wide range, from the qualitative to the quantitative, must take place in terms of invariancies derived from the properties of the underlying measurements. The lesson from abstract algebra and physics is that a group theoretical description is a useful tool in this respect. Hence, our iterated discussions aiming at the establishment of the group of transformation under which some property is invariant and at the corresponding group theoretical relationship between invariant properties, served a purpose above that of characterizing controllability and observability. Namely, to illustrate the power and the possibilities that modern control theory may accrue from including the group theoretical approach among its techniques.

Perhaps, by comparison of the discussions in part I and here it may seem that the tensor approach and the digraph approach have been introduced as alternative otherwise seemingly unrelated, representations of mathematically identical groups which define the structural or invariant properties of a system in a purely abstract way. Yet, it may be shown that the two different representations will complement each other in a manner which does make sense from the viewpoint of establishing qualitative design rules. An investigation of this problem will be undertaken in part III where we will show that system aggregation is the inherent aim of a tensor formulation, whereas system decomposition is the inborne intent of a digraph formulation. Thus, we shall demonstrate how, by extending the tensor approach to include output controllability, a new dual concept, called input observability, may be derived. Similarly, we shall show how determination of term rank may be facilitated for high order systems by a graph-theoretical decomposition of the system equations which has important implications also in a quantitative computational sense.

PART III: DIGRAPH DECOMPOSITION AND TENSOR AGGREGATION

INTRODUCTION

A formal comparison of Kalman's and Gilbert's criteria of controllability and observability has as its obvious result that the latter is a special case of the
former. Indeed, it is for this reason that Gilbert had to introduce the explicit assumption on the distinct eigenvalues. Yet, in a more pragmatic sense, this
is perhaps not the most important difference. Rather, from this point of view
the fundamental difference seems to be one of attitude. Thus, Kalman had the
general and all-inclusive case in mind, whereas Gilbert was oriented towards a
breaking down of the problem in order to simplify its solution. It is a striking
fact that even if the tensorial approach and the digraph approach are commonly
concerned with the structural properties of a system, the former aims at a compact description of the aggregated system, whereas the latter opens up for a
structural decomposition of the system into its component parts. It is not a
coincidence, therefore, that in the previous discussion of parts I and II Kalman's criterion was taken as the starting point of the tensorial approach,
while Gilbert's criterion was related to the digraph approach.

The purpose of the present part III is , based on an investigation of the structural properties of a system, to elucidate this difference in aim between the
tensorial and the digraph approach. Beginning with the latter the system digraph is submitted to a structural decomposition into its strong cyclic components and its condensed acyclic digraph. Essentially, considering the reachability matrix, this decomposition is achieved by a Boolean counterpart of the
wellknown quantitative operation of partitioning a matrix into its symmetrical
and antisymmetrical component matrices. The effect of this Boolean operation,
is that the system equations may be blocked into a hierarchy of independent subsets each of which may be dealt with separately. For high order systems this
may greatly simplify the solving of the equations and the finding of the eigenvalues. However, a more important perspective with particular relevance to
non-linear systems, may be that it establishes a close relationship between
distinct quantitative properties and well-defined structural characteristics.
Thus, by the digraph formulation we are led to a decomposition of a system
laying bare its characteristic properties.

A universal tensor which combines all the quantitative rank tests into a single
entity, may be formulated on the basis of an aggregation of the state and output equations. It is noteworthy that this tensor will encompass the additional
concept of output controllability in such way that, by symmetry considerations,
we are "forced" to introduce a new dual concept to be termed input observability. Turning to a digraph interpretation of this set of dual concepts their
structural counterparts are established and a defining pair of reachability and
term rank tests are formulated. Thus, by the tensor formulation we are led to
an aggregation that imposes upon all the inherent properties the symmetries valid
in general for any system.

Chapter 6. ON PARTITIONING OF A DIGRAPH

A system coefficient matrix is decomposable if it is possible to specify a permutation matrix which by a congruence transformation will bring the system equations into a block-triangular form. With a composite system in this hierarchical
form each block defines an independent subset of equations that may be dealt with
separately. It is evident that for high order systems determination of the term
rank or the eigenvalues, or solution of the system equations, is greatly facilitated if the system can be brought into this desirable form. Graph theory provides a well established scheme for bringing about the defining permutation matrix in terms of the socalled quasi-levels. This concept which furnishes a topological ordering of the vertices of a digraph, derives from a decomposition of
the digraph into its component cyclic and acyclic parts.

In view of the rich documentation of the theoretical aspects of this digraph
decomposition approach the emphasis in this section will be on the establishment
of an operational procedure for bringing about the desired form in an organized
manner. Though, we shall deviate on one important theoretical point. That is,
we shall base ourselves on the neglected fact that a decomposition of the digraph into its cyclic and acyclic component parts, is defined by a corresponding partitioning of its reachability matrix into a symmetrical and an antisymmetrical matrix. This partitioning forms a Boolean counterpart of the wellknown partitioning of a quantitative matrix into its symmetrical and antisymmetrical matrices. It follows that the important physical ramifications stem-
ming from the lastmentioned quantitative operation, may now be related to distinct structural properties of the system. Clearly, this may provide an ope-
ning towards a more fundamental treatment of non-linear systems. To indicate
the nature of such an approach we shall end this section by briefly considering
some aspects of the relationship between the system structure and the problem of
inverting sparse matrices.

6.1 The Structural Component Parts

To cope with systems of high dimensionality it may be advantageous to combine
the term rank test with a structural decomposition derived from the digraph representation of the system. Of course, to be feasible this decomposition procedure requires that the system is described by a square Boolean matrix that
is reducible. In practice, therefore, this approach may be considered as a sparse matrix technique assuming either that the system matrix has an overwhelming
number of zeros or that its zeros are strategically located (Harary, 1959).
In this role the decomposition procedure enjoys a certain popularity,particu-
larly within areas such as production engineering, management, and macro economics. An early application related to the control theory field seems to be
a chemical process simulator developed in Japan (Iri et al., 1971). However,
it is only quite recently, in connection with the decomposition of high order
systems, that this problem seems to have attracted more than a cursory attention in the control literature (Davison, 1977; Siljak, 1977a & b).

The graph-theoretical foundation of the decomposition method is well-established
in the textbook literature (Harary et al., 1965; Chrisofides, 1975). Very
briefly, the technique is founded on a determination of the socalled *quasi-
levels* of the vertices of the successor (predecessor) digraph. Based on these
quasi-levels the digraph is decomposed into its disjoint cyclic subsets, the
strong components, together with its aggregated acyclic subsets produced from
the *condensed acyclic digraph* by elimination of the "cyclic" vertices. The
corresponding decomposition of the defining matrix is brought about by a congruence transformation in which its row and its column indices are simultaneously
permuted as determined by the quasi-levels. Hence, if the matrix is reducible
its term rank may be determined by a term rank summation over the submatrices
specified by the abovementioned distinct component parts of the digraph. It is

noteworthy that the structural decomposition defined by the quasi-levels, is invariant under the symmetric group. It follows that, in the decomposition procedure, we are concerned with Boolean tensor properties valid in general for measurements on all scales.

By virtue of the fact that the procedure sketched above for structural decomposition may be said to be well documented, at least from a theoretical point of view, we shall confine ourselves here to the problem of making the approach operational. That is, we shall demonstrate how, in a few conceptual steps, it may be implemented in APL. However, before we enter upon this undertaking it will be worthwhile to clarify the concepts involved from a purely graph-theoretical point of view by considering a simple illustrative example.

To this end assume given the following, obviously sparse, Boolean matrix M:

$$\rho \square \leftarrow M$$

```
0 0 0 0 0 1 0 0 0 0
0 0 0 0 1 0 0 1 1 0
0 1 0 0 1 0 0 1 0 0
0 0 0 0 0 1 0 0 0 0
0 0 1 0 0 0 0 0 0 0
1 0 0 0 0 0 0 0 0 0
0 0 0 1 0 0 0 0 0 0
0 0 1 0 0 0 0 0 0 0
0 0 0 1 0 0 1 0 0 0
0 0 0 0 1 0 0 0 1 0
10 10
```

(1)

which, applying the APL function *ASSIGN* defined in part II, is found to have the term rank:

$$+/+/\underline{ASSIGN}\ M$$
8

(2)

Now, representing this matrix M by its conventional successor digraph of Fig. 1A, let us produce the structurally distinct component parts of this digraph in order to demonstrate that the term rank (2) may be found by a summation of the term ranks of these parts. First, we identify the strong components characterized by the fact that in each of them any pair of vertices are mutually reachable. Thus, by visual inspection of Fig. 1A the two strong components of Fig. 1B are established. It is easy to see that these two components, defined respectively by the two vertices 1 and 6 and the four vertices 2, 3, 5, and 8, are the only maximally strong, cyclic components of Fig. 1A. Secondly, we shrink each of these strong components into a single vertex. The outcome of this process which is known as the *condensation of a digraph with respect to its strong components*, is the socalled *condensed acyclic digraph* exhibited in Fig. 1C. Clearly, by the separation of the strong components of Fig. 1B from the condensed digraph of Fig. 1C, we have achieved a partition of the given digraph of Fig. 1A into two equivalence classes of respectively cyclic and acyclic digraphs. Though, in the latter class we still maintain a vertex representation of each of the strong components. These, let us call them, *strong vertices* are indicated by circles instead of crosses in the condensed acyclic digraph of Fig. 1C. Thirdly, therefore, to produce the distinct acyclic component contributing to the term rank of the given digraph, the strong vertices must be eliminated. The result of this operation is the aggregated acyclic digraph exhibited in Fig. 1D. Hence, in three, conceptually simple, steps we have structurally decomposed the successor digraph of Fig. 1A into its distinct component parts: the two strong components of Fig. 1B and the aggregated acyclic digraph of Fig. 1D.

Comparison of the strong component defined by the four vertices 2, 3, 5, and 8 in Fig. 1B with the cyclic digraph of Fig. II.6B*) and of the aggregated digraph of Fig. 1D with the acyclic digraph of Fig. II.6A, immediately reveals that we have encountered these component parts of the digraph of Fig. 1A before.   In fact, summing the unit entries of their respective maximal permutation matrices (II.40a&b), it is readily deduced that they each have a term rank of 3. It is not difficult to see, either by inspection or by applying the alternating path method, that the strong component defined by the pair of vertices 1 and 6 in Fig. 1B, must have the term rank 2.   Thus, adding up the term ranks of the three distinct components depicted in Figs. 1B & C we find a total of 8 in agreement with (2).

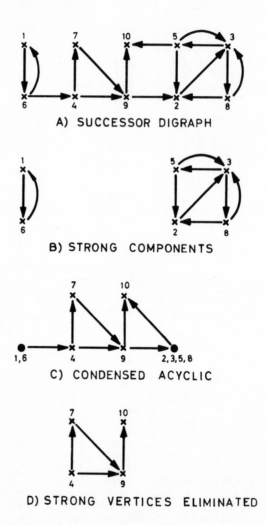

Fig. 1.   A Topological Decomposition

*) A reference number preceeded by a Roman numeral I or II refers to the item of that number in part I or part II respectively.

6.2 Introducing Level Coding

To implement this approach in APL we must substitute the visual inspection of the given digraph by an organized set of operations on a suitable matrix representation of the digraph. For the latter purpose we choose to introduce, for reasons which later will become apparent, the reachability matrix R. Basically, this matrix may be determined applying the APL function _REACHABILITY_, defined in part II, to the matrix M of (1). In practice, however, it may be desirable to revise this function to make it more efficient. A simple improvement, indicating what can be done in this direction, is introduced in the following variation of the algorithm called _REACHABILITY_1:

```
        ∇ REACHABILITY1 [□] ∇

        ∇ R←REACHABILITY1 M;AUX
[1]     ⍝
[2]     ⍝    THE REACHABILITY MATRIX ¨R¨ IS
[3]     ⍝    ESTABLISHED ITERATIVELY FROM A
[4]     ⍝    SQUARE BOOLEAN MATRIX ¨M¨ SUCH THAT
[5]     ⍝    AT STEP N ¨M¨ HAS THE POWER 2*N.
[6]     ⍝
[7]     INITIAL:R←M∨(⍳1↑⍴M)∘.=⍳1↑⍴M
[8]     NEXT:AUX←R
[9]     R←R∨R∧.∧R
[10]    →(∨/∨/AUX≠R)/NEXT
        ∇
```

The efficiency of this function, which is valid in either origin, derives from the formulation of statement [9]. Thus, in comparison with statement [7] of the function _REACHABILITY_ defined in part II, it is readily seen that it progresses through the Boolean power series, specifying R, by M to the power 2^N instead of by M to the power N with N being the number of iterations.

Applying the APL function thus established to the successor matrix M of (1) the reachability matrix R is determined as follows:

$$\rho□←R←\underline{REACHABILITY}1\ M$$

```
1 0 0 0 0 1 0 0 0 0
1 1 1 1 1 1 1 1 1 0
1 1 1 1 1 1 1 1 1 0
1 0 0 1 0 1 0 0 0 0
1 1 1 1 1 1 1 1 1 0
1 0 0 0 0 1 0 0 0 0
1 0 0 1 0 1 1 0 0 0
1 1 1 1 1 1 1 1 1 0
1 0 0 1 0 1 1 0 1 0
1 1 1 1 1 1 1 1 1 1
10 10
```

(3)

The significance of introducing the reachability matrix R instead of the successor matrix M to represent the digraph, is, as pointed out earlier, that R implements all the logical consequences of the transitive law (Franksen,1977). In fact, if, in the digraph of Fig. 1A, a vertex J is reachable from a vertex I over a topological distance larger than one, R augments the digraph by an oriented edge from I to J. We should not expect, therefore, a direct correspondence between the approach given in Fig. 1 and the approach we are now going to undertake in terms of the reachability matrix R. Though, in terms of fundamental principles the two approaches do not differ.

The immediate purpose of the operations to which matrix R will be submitted, is to obtain a description of the digraph in terms of a structural or topological ordering of its vertices. The process by means of which the vertices are ordered is known as *level coding*. Essentially, the *level* of a vertex is a nonnegative integer which, assigned to the vertex, indicates its topological distance along some directed path in the digraph from the set of sources or, conversely, to the set of sinks. By virtue of the fact that a vertex is reachable from itself along a directed path containing a zero number of directed edges the sources or, conversely, the sinks are assigned zero level. It is readily appreciated that level assignment presupposes the existence of sources or sinks. Furthermore, apart from sources or sinks, level assignment is not unique since clearly it may be possible to assign more than one level to an arbitrary vertex.

To guarantee the existence of at least one source and at least one sink it is taken as a basic assumption of any level assignment that the digraph is *acyclic*. Hence, the concept of levels has meaning only in relation to the vertices of an acyclic digraph. Granting the satisfaction of this assumption a unique level assignment may be defined, say, by the directed paths of *maximal distance*. A level assignment based on this rule is called *low-level coding* if the sources are assigned zero level and *inverted-level coding* if the sinks are assigned zero level (Dzubak & Warburton, 1965). The term "inverted" refers to the fact that the two kinds of level assignments are directional duals. That is, submitting the converse digraph to low-level coding is tantamount to applying inverted-level coding to the original digraph. Yet, it should be noted that this duality is related solely to the arguments of the operation and not necessarily to its resulting level assignments. Therefore, a set of vertices assigned the same level by low-level coding need not be assigned the same level by inverted-level coding. The reason for this is, of course, that even if they are placed at equal maximal distance from the sources, this does not imply that they are placed also at equal maximal distance from the sinks.

Level assignment on the vertices of an acyclic digraph by low-level or inverted-level coding may be performed iteratively by operations on the square Boolean matrix M representing the digraph (Harary et al., 1965). By virtue of the fact that the converse digraph is represented by the transpose of M, it will suffice to consider low-level coding on M conceived as a predecessor matrix. Since the digraph is acyclic it will have one or more sources each represented by a zero column in M. Hence, to begin with, the vertices, identified by the zero columns of M, are assigned level zero. These vertices are now eliminated. That is, the said columns and the corresponding rows are deleted in M. The remaining submatrix of M represents an acyclic subgraph the sources of which are the vertices at topological distance one from the sources of the original digraph. The set of vertices thus identified as sources by the zero columns of the remaining submatrix, are assigned level one and the said columns and the corresponding rows are deleted producing a new submatrix. Step by step this process of level assignment and source elimination is continued until the entire matrix M is elided. Evidently, this will occur in a finite number of steps ending up with the last level assignment specifying the distance of the longest path in the digraph.

An APL function *LEVELCODE*, implementing this procedure in either origin, is the following:

```
     ∇ LEVELCODE [□] ∇

     ∇ L←LEVELCODE M;B;AUX
[1]   ⍝
[2]   ⍝      ¨M¨ IS THE PREDECESSOR MATRIX FOR
[3]   ⍝      LOW-LEVEL CODING AND THE SUCCESSOR
[4]   ⍝      MATRIX FOR INVERTED-LEVEL CODING.
[5]   ⍝      EMPTY LEVEL ASSIGNMENT IF CYCLIC.
[6]   ⍝
[7]   INITIAL:L←(1↑ρM)ρ0
[8]    B←∨/[□IO] M
[9]   NEXTΔSUBGRAPH:L←L+B
[10]  END:→0×ι~∨/B
[11]   AUX←B
[12]   B←B∨.∧M
[13]   →(∨/AUX≠B)/NEXTΔSUBGRAPH
[14]  CYCLICΔGRAPH:L←ι0
     ∇
```

In this function the level assignments of the digraph vertices are specified by the vector L. The Boolean vector B identifies in each step the remaining vertices. That is, the unit entries of B define the columns of M not yet eliminated. The introduction of a vector B for this purpose has several advantages. First, in [9] it gives rise to a simple updating of the level assignments. Secondly, in [10] it is used to check whether all vertices have been eliminated. Thirdly, in [12] elimination of the rows, corresponding to the identified sources are accomplished in a single inner product. And, fourthly, in [13] cyclic components, contrary to the basic assumption, are revealed. Indeed, if the basic assumption requiring an acyclic digraph is violated, the level assignment in [14] will produce an *empty vector of levels*. The advantage of introducing this special level definition for cyclic digraphs, is that the concept of level assignment is now made operational for digraphs in general. To illustrate, a useful consequence of this generalization is that the level concept, implemented in a function like *LEVELCODE*, may be used also to determine whether a digraph is cyclic or acyclic.

At this point it is obvious, since we are concerned with cyclic digraphs like the successor digraph of Fig. 1A, that the object of a non-trivial level assignment must be to accomplish a partition of the digraph, as illustrated in Figs. 1B & C, into its cyclic and acyclic component parts. It was for this reason that matrix R instead of matrix M was adopted to represent the digraph. In fact, the reason that we thus deviate from the conventional approach, is that only R will provide, at least in principle, a direct solution to this problem.

In conventional matrix algebra it is wellknown that any square matrix N may be written as a sum of a *symmetrical* matrix:

$$0.5 \times (N + \lozenge N) \tag{4a}$$

and an *antisymmetrical* matrix:

$$0.5 + (N - \lozenge N) \tag{4b}$$

Less wellknown, perhaps, is it that in Boolean algebra any square Boolean matrix M may be similarly partitioned (Luce, 1952; Franksen, 1976). Thus, defining Boolean matrix addition in terms of the connective disjunction, M may be written as the Boolean "sum" of a *symmetrical* matrix:

$$M \wedge \lozenge M \tag{5a}$$

and an *antisymmetrical* matrix:

$$M \wedge \sim \mathbb{Q}M \tag{5b}$$

It is not difficult to see that if M is the successor (predecessor) matrix of some digraph, then the symmetrical matrix (5a) will represent the *cyclic* components at topological distance one of the digraph. Similarly, the antisymmetrical matrix (5b) will represent what, at topological distance one, appear to be the *acyclic* components of the digraph. An acyclic digraph, therefore, is characterized by the fact that (5a) is a null matrix for all non-zero Boolean powers of M.

The observation that the symmetrical and antisymmetrical Boolean matrices (5) will identify the cyclic and the acyclic components respectively at topological distance one of the digraph defined by matrix M, is of particular interest if M is substituted by the reachability matrix R. The characteristic feature of R, is that any consequence of the transitive law is represented by a single edge. That is, R depicts by directed paths of topological distance one all the cyclic and acyclic components of the digraph in question. It follows that a partition of R by (5) will disclose these components, and when submitting R to (5a), it is evident that the symmetrical component matrix:

$$R \wedge \mathbb{Q}R \tag{6a}$$

will identify all the strong components by virtue of the fact that mutual reachability will exist between every pair of vertices in a strong component. Though, it should be noted, recalling the Boolean tensor polynominal Q from which R is defined (II.3), that in addition to the strong components of the given digraph each vertex by itself is considered to be a strong component. That is, the definition of R may be said to introduce a loop for each vertex representing the reflexive property that each vertex may be compared with itself. The algebraic implication of this is that the trace of R, and hence of (6a), consists of unit entries only.

This generalization, imparted to (6a) by the definition of R, must be seen in contradistinction to the purely acyclic properties of the antisymmetrical component matrix:

$$R \wedge \sim \mathbb{Q}R \tag{6b}$$

obtained from (5b). To be sure, the antisymmetrical matrix (6b) represents but the acyclic properties of the given digraph. It follows that we may submit (6b) to a level assignment the result of which, relative to the original cyclic digraph, may be given an interpretation, say, as *generalized levels*. In other words, the level concept may be generalized from acyclic to cyclic digraphs simply by performing the level coding on the antisymmetrical component of the reachability matrix of the digraph in question. Clearly, this generalization includes acyclic digraphs as the special case in which the symmetrical component of the reachability matrix vanishes except for the unit matrix referred to previously.

To illustrate, invoking the function *LEVELCODE* on (6b) with the reachability matrix defined by (3), produces by inverted-level coding the following vector of generalized levels for the vertices of the cyclic digraph of Fig. 1A:

$$\rho \square \leftarrow \underline{LEVELCODE} \ R \wedge \sim \mathbb{Q}R$$

5 1 1 4 1 5 3 1 2 0
10
$$\tag{7}$$

Now, by comparison with Fig. 1B it is immediately seen that all the vertices defining a strong component are assigned the same level. Obviously, this is a general property. We may say, therefore, that the generalized levels introduce a topological ordering of the vertices that partitions the digraph into its component cyclic and acyclic parts. Though, it is evident that this partitioning

is not in general unique. Thus, even if all vertices defining a strong component are assigned always the same generalized level, it may happen that this level is assigned also to vertices not belonging to the strong component in question. Hence, to obtain a unique partitioning of the digraph vertices some transformation is needed of the generalized levels such that only vertices belonging
to the same strong component are assigned the same level.

With this in mind let us relate the generalized levels of (7) to the vertices
of the condensed acyclic digraph of Fig. 1C. By this comparison it is important to observe that no two vertices of the condensed digraph are assigned the
same level. Of course, in the present case it it a happy coincidence. But,
taken together with the fact that all vertices defining a strong component are
assigned the same level, it has the obvious consequence that only vertices belonging to the same strong component are assigned the same level. A level assignment of a cyclic digraph satisfying these properties, is called a *quasi-
level* assignment (Harary et al., 1965). What is needed, therefore, is a transformation that turns the generalized levels into quasi-levels.

6.3 Topological Ordering by Quasi-Levels

From an operational point of view the defining property of a quasi-level assignment of a cyclic digraph, may be characterized by the fact that each vertex of
the condensed acyclic digraph is assigned a distinct level. It is wellknown
that any level assignment by low-level or inverted-level coding of an acyclic
digraph, submits to a transformation such that no two vertices have the same
level (Harary et al., 1965). Hence, the quasi-levels may be determined by a
level assignment on the condensed acyclic digraph followed by a transformation
of this assignment into a corresponding set of distinct vertex levels. The
problem, therefore, of turning generalized levels into quasi-levels, reduces to
to the far simpler problem of establishing a matrix representation of the condensed acyclic digraph. The solution to this problem is founded, as we shall
see, on the determination of the symmetrical component of the reachability matrix.

Considering the derivation of a condensed acyclic digraph from a given cyclic
digraph as a transformation, in fact a surjection, it is noteworthy that the
transformation matrix, say W, may be established from the symmetrical component
matrix (6a). The procedure by means of which this is done, is implemented in
the following APL function, *TRANSFORM*, that is valid in either origin:

$$\triangledown \; \underline{\mathit{TRANSFORM}} \; [\Box] \; \triangledown$$

```
       ∇ W←TRANSFORM R
[1]    ⍝
[2]    ⍝   THE TRANSFORMATION MATRIX "W" FROM A CYCLIC
[3]    ⍝   GRAPH REPRESENTED BY ITS REACHABILITY MATRIX
[4]    ⍝   "R" TO ITS CONDENSED GRAPH IS FOUND.
[5]    ⍝
[6]    SYMMETRICAL:W←<\R∧⍉R
[7]    W←(∨≠W)/W
       ∇
```

Applying this function to the reachability matrix (3), produces the matrix W
that transforms the successor digraph of Fig. 1A to the condensed acyclic digraph of Fig. 1C:

$$\rho \square \leftarrow W \leftarrow \underline{TRANSFORM} \; R$$

```
1 0 0 0 0 0
0 1 0 0 0 0
0 1 0 0 0 0
0 0 1 0 0 0
0 1 0 0 0 0
1 0 0 0 0 0
0 0 0 1 0 0
0 1 0 0 0 0
0 0 0 0 1 0
0 0 0 0 0 1
10 6
```

(8)

The row indices of this matrix specify the vertices of the given cyclic digraph, whereas the column indices identify the vertices of the condensed acyclic digraph. The characteristic property of matrix W, therefore, is that the unit entries of each column define the set of vertices making up the strong component in question.

The transformation from the Boolean matrix M of (1), defining the successor digraph of Fig. 1A, to the matrix, specifying the condensed digraph of Fig. 1C but with a loop assigned to each of the two strong vertices, is given by the expression:

$$\rho \square \leftarrow (\lozenge W) \vee . \wedge M \vee . \wedge W$$

```
1 0 0 0 0 0
0 1 0 0 1 0
1 0 0 0 0 0
0 0 1 0 0 0
0 0 1 1 0 0
0 1 0 0 1 0
6 6
```

(9a)

However, it is more convenient to submit the antisymmetrical component (6b) of the reachability matrix to this transformation, since the outcome of this operation is simply the reachability matrix of the condensed acyclic digraph of Fig. 1C without any additional loops assigned to the vertices.

An APL function, *CONDENSED*, that implements this procedure independent of origin, may be the following:

```
     ∇ CONDENSED [□] ∇

     ∇ P←W CONDENSED R
[1]   ⋀
[2]   ⋀      DETERMINATION OF THE REACHABILITY MATRIX
[3]   ⋀      "P" OF A CONDENSED ACYCLIC DIGRAPH GIVEN
[4]   ⋀      THE CORRESPONDING CYCLIC DIGRAPH MATRIX
[5]   ⋀      "R" AND THE ASSOCIATED TRANSFORMATION
[6]   ⋀      MATRIX "W".
[7]   ⋀
[8]   ANTISYMMETRICAL:P←R∧⍉~R
[9]   P←(⍉W)∨.∧PV.∧W
     ∇
```

Invoking this function with arguments W and R specified by (8) and (3) respectively, the following reachability matrix RC is determined for the associated condensed acyclic digraph:

$$\rho\square\leftarrow RC\leftarrow W \underline{\textit{CONDENSED}} R$$

```
0 0 0 0 0 0
1 0 1 1 1 0
1 0 0 0 0 0
1 0 1 0 0 0        (9b)
1 0 1 1 0 0
1 1 1 1 1 0
6 6
```

Performing in 1-origin a level assignment on this matric RC submitting it to the function *LEVELCODE*, yields:

$$\rho\square\leftarrow I\leftarrow\underline{\textit{LEVELCODE}}\ RC$$

```
5 1 4 3 2 0          (10a)
6
```

This level assignment I on the condensed acyclic reachability matrix RC, may be made distinct relative to the vertices of the condensed digraph submitting I to a repeated grade-up operation. The outcome of this operation, $\Delta\Delta I$, transformed to the cyclic digraph by the transformation matrix W, will produce the quasi-levels, say J, of that digraph:

$$\rho\square\leftarrow J\leftarrow W+.\times\Delta\Delta I$$

```
6 2 2 5 2 6 4 2 3 1   (10b)
10
```

Apart from a constant difference of a unit value on all elements this result is seen to agree with (7). In general, it should be noted that level assignment is an order relation and, hence, invariant under the socalled monotonic group (Franksen, 1975). We should not expect, therefore, that level assignment is unique in any quantitative sense.

Previously, it has been stated that the quasi-levels J define a topological order which partitions a cyclic digraph into its component cyclic and acyclic parts. The former, of course, are the strong components. Essentially, this topological order is determined by a permutation which may be produced from J of (10b) by, say, a grade-down operation yielding for 1-origin:

$$\rho\square\leftarrow\Psi J$$

```
1 6 4 7 9 2 3 5 8 10   (11)
10
```

Submitting the row and column indices of the Boolean matrix M of (1) to this permutation, transforms M into the following lower block-triangular form:

$$M[\Psi J;\Psi J]$$

```
0 1|0 0 0 0 0 0 0 0
1 0|0 0 0 0 0 0 0 0
0 1|0|0 0 0 0 0 0 0
0 0 1|0|0 0 0 0 0 0
0 0 1 1|0|0 0 0 0 0     (12)
0 0 0 0 1|0 0 1 1|0
0 0 0 0 0|1 0 1 1|0
0 0 0 0 0|0 1 0 0|0
0 0 0 0 0|0 1 0 0|0
0 0 0 0 1 0 0 1 0|0
```

Considering this matrix the reader will immediately recognize six blocks: two submatrices representing each a strong component of 2 and 4 vertices respectively (cf. Fig. 1B); and four single-element submatrices representing the acyclic components (cf. Fig. 1D). Obviously, the number of blocks corresponds to the total number of vertices in the condensed acyclic digraph (cf. Fig. 1C) just as the number of elements in (12) corresponds to the total number of vertices in the original cyclic digraph. In APL we may use the transformation matrix W of (8) and the condensed acyclic level assignment I of (10a) to determine the total number of elements (or cyclic digraph vertices) in each block. Thus, in 1-origin we find:

$$\rho\square\leftarrow B\leftarrow(+/W)[\Psi I]$$

2 1 1 1 4 1

6

(13)

Perhaps, the most significant characteristic of the block-triangular form of (13), is that each strong component is represented by a distinct submatrix whereas, in contradistinction, the acyclic component part is depicted in general as a set of disjoint submatrices. The term rank determination, however, requires that the acyclic component part is treated as a single submatrix. Hence, it will be necessary to submit (12) to some further partitioning operations.

Though, before we proceed with this problem, it may be worthwhile to point out that (11) is a convenient way of expressing the permutation matrix P:

$$\rho\square\leftarrow P\leftarrow((\iota 1\uparrow\rho M)\circ.=\iota^{-}1\uparrow\rho M)[;\Psi J]$$

```
1 0 0 0 0 0 0 0 0 0
0 0 0 0 0 1 0 0 0 0
0 0 0 0 0 0 1 0 0 0
0 0 1 0 0 0 0 0 0 0
0 0 0 0 0 0 0 1 0 0
0 1 0 0 0 0 0 0 0 0
0 0 0 1 0 0 0 0 0 0
0 0 0 0 0 0 0 0 1 0
0 0 0 0 1 0 0 0 0 0
0 0 0 0 0 0 0 0 0 1
10 10
```

(14)

Thus, in terms of this matrix P the permutation operation may be rewritten:

$$(\lozenge P)\vee.\wedge M\vee.\wedge P$$

(15a)

or, in general, if M is a quantitative matrix:

$$(\lozenge P)+.\times M+.\times P$$

(15b)

The existence of the permutational transformation (15) lies at the foundation of the theoretical proofs given previously on the determination of potential controllability (Shields & Pearson, 1976; Glover & Silverman, 1976). The expression (12) with the permutation defined in terms of the quasi-levels J, provides an operational counterpart to these proofs. In this connection it may be of interest to compare the form of the transformation (15a) with that of the alternating path transformation (II.43c). The difference in form observed here, reflects the fact that (15a) is invariant under the symmetric group, whereas (II.43c) is invariant only under the alternating group.

But let us turn back to the problem of determining the acyclic component part of (12). Considering the cyclic component (6a) of the reachability matrix R it is easy to see that we may determine the vertices defining strong components by the expression:

$$1 < +/R \wedge \Diamond R \tag{16}$$

However, by this expression we may have eliminated also such vertices which define loops. Information about this special kind of strong components is given by the unit entries in the trace of the defining Boolean matrix M. Adding this information to that of (16) yields in 1-origin:

$$(1 \ 1 \Diamond M) \vee 1 < +/R \wedge \Diamond R \tag{17a}$$

Hence, with M and R given by (1) and (3) respectively a Boolean vector V, specifying by its unit entries the acyclic vertices of M, may be established as follows:

$$\rho \Box \leftarrow V \leftarrow \sim (1 \ 1 \Diamond M) \vee 1 < +/R \wedge \Diamond R$$

```
0  0  0  1  0  0  1  0  1  1
10
```
(17b)

Submitting the permuted matrix M of (12) to a transformation, in fact an injection, by V similarly permuted, yields a single submatrix defining the acyclic component part depicted in Fig. 1D:

$$V[\Psi J]/V[\Psi J] \neq M[\Psi J; \Psi J]$$

```
0  0  0  0
1  0  0  0
1  1  0  0
0  0  1  0
```
(18)

Clearly, this matrix is identical with the matrix M1 of (II.39a) the permutation matrix (II.40a) of which determined its term rank to 3.

In a similar manner, introducing the negation of V, the matrix representation of the two strong components of Fig. 1B is produced:

$$(\sim V[\Psi J])/(\sim V[\Psi J]) \neq M[\Psi J; \Psi J]$$

```
0  1 | 0  0  0  0
1  0 | 0  0  0  0
-----+-----------
0  0 | 0  0  1  1
0  0 | 1  0  1  1
0  0 | 0  1  0  0
0  0 | 0  1  0  0
```
(19)

This matrix, of course, may be further partitioned into its two non-zero submatrices. The largest of these submatrices may be identified with the matrix M2 of (II.39b) whose permutation matrix (II.40b) was found to yield the term rank 3. By inspection it is evident that the term rank of the smallest submatrix of (19) is 2.

Summing up, therefore, we see that the expressions (18) and (19) establish three submatrices the term rank of which may be determined independently of each other such that, when added, they produce the total term rank (2) of the defining matrix M of (1).

6.4 On Decomposable Systems

Decomposition of a square matrix M by a congruence transformation defined by
its quasi-levels J, is a structural operation invariant under the symmetric
group. Hence, it applies to any square matrix whether its entries are Boolean
or have some quantitative values. Of course, the usefulness of this permutational operation is confined to decomposable, usually sparse, matrices. In the af-
firmative case, assuming M to be a quantitative square matrix, the decomposition
procedure is essentially a question of topological reordering. Based on the APL
functions developed in the previous section this procedure is summarized under
the heading: "Topological Reordering" in the upper half of Table 1.

MATRIX DECOMPOSITION	
TOPOLOGICAL REORDERING	$R \leftarrow \underline{REACHABILITY}1 \ M \neq 0$ $W \leftarrow \underline{TRANSFORM} \ R$ $RC \leftarrow W \ \underline{CONDENSED} \ R$ $I \leftarrow \underline{LEVELCODE} \ RC$ $J \leftarrow W + . \times \blacktriangle \blacktriangle I$ $B \leftarrow (+\neq W)[\Psi I]$
INVERSION	$N \leftarrow B \ \underline{INVLOW} \ M[\Psi J; \Psi J]$ $N[\blacktriangle \Psi J; \blacktriangle \Psi J]$

Table 1. Decomposition by Quasi-Levels

The computational advantage that may be gained from bringing a matrix into its
structurally decomposed form, goes beyond the term rank determination aimed at
in the previous section. In fact, if the decomposable matrix is quantitative
this approach will greatly facilitate the determination of its determinant, its
inverse, and its eigenvalues (Harary, 1959 & 1962). Furthermore, the structurally decomposed form provides the stepping stone to other and more elaborate
solution methods of simultaneous systems of equations (Steward, 1962).

To briefly illustrate these remarks let us consider the inversion of a decomposable matrix which, by the approach of Table 1, can be reordered into a lower,
quasi-triangular form with a blocking around the diagonal. The APL function
INVLOW implements such a matrix inversion procedure. The arguments of this function are the topologically reordered matrix M (i.e., a matrix of the form
$M[\Psi J; \Psi J]$ illustrated in (12)) and a blocking vector B (determined as demonstrated in (13)). The first APL expression in the lower half of Table 1, headed "Inversion", depicts the corresponding call invoking the function.

```
       ∇ INVLOW [□] ∇

       ∇ R←B INVLOW M;N;□IO;J;L
[1]   ⍝
[2]   ⍝      INVERSION OF A LOWER QUASI-TRIANGULAR
[3]   ⍝      MATRIX "M" WITH A BLOCKING AROUND THE
[4]   ⍝      DIAGONAL GIVEN BY VECTOR "B".
[5]   ⍝
[6]    INITIAL:□IO←J←1
[7]     R←(⍳1↑⍴M)∘.=⍳1↑⍴M
[8]     N←R-M
[9]     L←0
[10]  NXT△BLOCK:L←(⁻1↑L)+⍳(,B)[J]
[11]    R[L;]←R[L;]+(⌹M[L;L])+.×N[L;]+.×R
[12]    J←J+1
[13]   →(J≤⍴B)/NXT△BLOCK
       ∇
```

The fundamental principle underlying the matrix inversion procedure of this function, is formulated in statement [11]. Thus, by this statement we determine in each iteration step the entries of those rows L of the inverted matrix R that correspond to the complete block inversion ⌹M[L;L] of that step. The order in which we iterate from block to block is given by the entries of the blocking vector B. Accordingly, the iteration process proceeds corresponding to the levels of the condensed acyclic digraph. Hence, in each iteration we first calculate the entries N[L;]+.×R which determine the corresponding acyclic component part of the digraph. Multiplying the outcome of this calculation by the inverse matrix ⌹M[L;L] takes care of the strong component representing the influence upon this iteration step of the cyclic component part of the digraph. Finally, adding the combined contribution of the digraph involved in this step to the inverted matrix R[L;] of the previous step, updates the resulting inverse matrix R.

Clearly, the inverse matrix R resulting from this approach will be ordered according to the blocking of M. Hence, to rearrange R into the original order of matrix M, its rows and columns must be sorted performing a grade-up followed by a grade-down: ⍋⍒J, as illustrated by the last expression of Table 1. Of course, the grade-down operation in the two APL expressions in the lower half of Table 1 could equally well have been substituted by the grade-up operation. The significance of this substitution is a change from a lower to an upper quasi-triangular form of the matrix. To evade a corresponding revision of the function INVLOW we may simply transpose the matrix argument M before invoking the function.

Leaving the problem of matrix inversion, it should be emphasized that the consequences of bringing a matrix into its structurally decomposed form, reach beyond that of computational advantage. Thus, the perspective of the topological decomposition, is that it is the structurally defined properties that will carry over into the non-linear domain. For example, the relationship imposed by the digraph upon the eigenvalues could be such structural properties that remain invariant in the transition from linear to non-linear systems. From this viewpoint the procedure of term rank determination in combination with structural decomposition is of particular importance. Because by this approach we may not only find the structurally determined number of zero eigenvalues, but, even more, we may relate these zero eigenvalues to distinct component parts of the system structure. Thus, it is readily seen that the structurally determined zero eigenvalues fall into two separate classes. One class is defined applying the term rank test to each of the strong components. This may be called the *cyclic class* of zero eigenvalues. The members of the other, say, *acyclic class* of zero eigenvalues are specified by the combined acyclic subset in a number corresponding to the dimension of that subset. This relationship appears to be significant because it ties up the eigenvalue concept with the structural framework in which it is imbedded. Though, a discussion of this topic is outside the scope of this book.

Undoubtedly, the main advantage of a tensor formulation, is that it offers the most compact, aggregated description of any system in terms of its basic invariant characteristics. In this chapter the total system will be formulated in terms of a universal tensor polynomial which generalizes the tensor Q, introduced in part I, by being derived from a compounded matrix that gives an aggregated description of the state and output equation coefficient matrices. Curiously enough, this compounded matrix has been introduced recently for the opposite purpose of decomposition in order to obtain a Boolean analogy of Kalman's canonical structure of linear dynamic systems (Siljak, 1976a&b). The important aspect of the approach adopted here, however, is that the criteria of state controllability and observability are derived as simple symmetry properties of a single universal tensor. Simultaneously, though, the universal tensor also exhibits a distinct asymmetry that may be identified with the criterion of *output controllability* defined in chapter 2 section 2.1.

The conception of symmetry in science means that a description remains unaltered by any change in the reference system. Indeed, *symmetry in the scientific sense of invariance of algebraic forms,* becomes the expression of general conformity to basic laws and an indication of the intrinsic structure underlying the observed fundamental relationships. The importance of maintaining algebraic symmetry as a formulation technique to depict real world invariance, is substantiated by the common experience that an apparent asymmetry most often will reveal a hidden fact which had remained unobserved hitherto. In general, the procedure of identifying the nature of a neglected fact, is to modify the description of reality in such a way that by taking this fact into account *symmetry will be restored again* in the formulation.

The asymmetry introduced into the formulation of the universal tensor by the identification of the concept of output controllability, indicates that to restore the symmetry a new concept, called *input observability*, must be defined (Franksen et al., 1976). The structural properties of the new pair of dual concepts, output controllability and input observability, are derived by analogy with the qualitative design concepts of potential state controllability and potential observability. In particular, it will be shown that, even if both pairs of dual concepts originate in the reachability properties of the same digraph, they are concerned with entirely different aspects of these structural properties. This difference is further emphasized by the fact that the new pair of dual concepts, requires the establishment of an entirely different formulation of the term rank tests. A formulation which, based on the digraph decomposition concept, requires us to add the term ranks of two independent subsets of coefficient matrices. Another feature of this discussion of output controllability and input observability, is the fact that it permits us to touch upon some of the inherent possibilities of using a qualitative approach as a design tool in the area of automatic control.

7.1 Establishing a Universal Tensor

For reasons that will shortly become evident we shall aggregate the state and output equations (I.1 & 2) into a single matrix equation which in conventional mathematical notation may be written:

$$
\begin{bmatrix} Y \\ \dot{X} \\ \dot{U} \end{bmatrix} = \begin{bmatrix} 0 & C & D \\ 0 & A & B \\ 0 & 0 & 0 \end{bmatrix} \times \begin{bmatrix} \dot{Y} \\ X \\ U \end{bmatrix} \tag{20}
$$

Consider in this expression the *compounded system coefficient matrix* S:

$$S = \begin{bmatrix} 0 & C & D \\ 0 & A & B \\ 0 & 0 & 0 \end{bmatrix} \tag{21}$$

Obviously, if we delete from this matrix the row and column (corresponding to the output vector Y and its time-derivative) we end up with the augmented-system matrix F of (II.56). Similarly, if we delete the *last* row and column (corresponding to the input or control vector U and its time-derivative) we produce the coefficient matrix H of (II.72) dual to F. Thus, the compounded system matrix S attains some of its importance by exhibiting the dual matrices F and H as a symmetrical pair of submatrices. However, we may attach importance beyond that to the symmetrical form of S. In particular, we see that the zero row and the zero column of S preserve the invariant properties; discussed in part II, that the time-derivative of the input or control vector U is not an explicit function of X or U (II.55) and that the time-derivative of the output vector Y appears as an ignorable variable. It is noteworthy that the aggregated expression (20) derives its innermost significance from these fundamental, yet implicit, assumptions of the format in which the state and output equations are traditionally cast.

From the specification (21) it is obvious that S is a square matrix of dimensions:

$$((\underline{Y}+\underline{X}+\underline{U}),\underline{Y}+\underline{X}+\underline{U}) = \rho S \tag{22}$$

It is apparent therefore, that, by analogy with the establishment by (I.12) of the Cartesian tensor polynomial Q from the coefficient matrix A, we may use the compounded system matrix S to derive for the aggregated system a corresponding Cartesian tensor polynomial, say, Z. By virtue of the fact that Z represents our universe of discourse we shall call Z the *universal tensor*. Assuming 1-origin the component array of Z is determined by the expression:

$$Z \leftarrow S0,[2]S1,[2]....,[1.5]SP \tag{23}$$

where the J'th power of S, invoking the APL function *PLUSAMPLY* defined in part I, is found in general by analogy with (I.9) as:

$$SJ \leftarrow S \ PLUSAMPLY \ J \tag{24}$$

For reasons of comparison it will be of interest to express the J'th power of S in conventional mathematical notation:

$$SJ = \left[\begin{array}{c|c|c} 0 & C \times A^{J-1} & C \times A^{J-2} \times B \\ \hline 0 & A^J & A^{J-1} \times B \\ \hline 0 & 0 & 0 \end{array} \right] \tag{25}$$

In this connection it will be noted that the term $S0$ in (23) is the 0'th power of S. Hence, $S0$ is a unit matrix.

Letting $\underline{P}$ denote the total number of powers of S, it follows from (22) that the dimensions of the component array of Z, determined by (23), are:

$$((\underline{Y}+\underline{X}+\underline{U}),\underline{P},\underline{Y}+\underline{X}+\underline{U}) = \rho Z \tag{26}$$

It is readily seen that if P is the highest power of S in (23), then the dimension $\underline{P}$ in (26) must be:

$$\underline{P} \leftarrow P+1 \tag{27}$$

However, this does not delimit this dimension of Z since in principle this tensor polynomial may be extended to any arbitrary power P of S. What is needed, therefore, is some relationship between the highest power of S in Z and the highest power of A in Q, so that we may utilize that the latter has been established already in (I.3).

In part II the digraph representation of the augmented-system matrix F of (II.67) was given in Fig. II.8, while the digraph representation of the dual matrix H of (II.77) was given in Fig. II.9. In these illustrations the matrices F and H were specified in terms of the Boolean matrices A, B, and C defined respectively by (I.56), (I.66), and (I.83). Based on these specifications, assuming D to be a null matrix, we may combine the two digraph representations of Figs. II.8 & 9 into a single successor digraph which, illustrating the compounded system matrix S of the form (21), is depicted in Fig. 2. Comparison of this aggregated successor digraph with the system successor digraph of Fig. II.1,

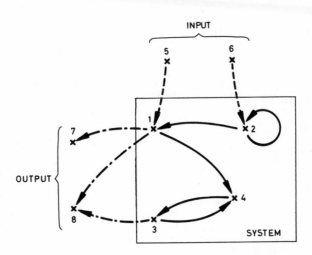

Fig. 2. An Aggregated Successor Digraph

brings out the interesting observation, valid in general, that to establish a reachability matrix for the former *two additional power terms* are required relative to the number of power terms necessary to establish the reachability matrix for the latter. With $\underline{X}$ designating the number of state variables the highest power of A in the tensor polynomial Q was specified by (I.3) to be $\underline{X}$-1. We would expect, therefore, that the highest power P of S, needed in (23) to establish the universal tensor Z, will be:

$$P \leftarrow \underline{X}+1 \tag{28}$$

Assuming this to be true, it follows by (27) and (26) that the shape of array Z is a rectangular box in contradistinction to Q the shape of which, we recall, is a cube. Though, we have as yet not confirmed that the dimension $\underline{P}$ of Z is determined by (27) combined with (28).

The fundamental significance of the universal tensor polynomial Z, is that in a single expression it describes simultaneously all the known criteria of controllability and observability. Truly, like the digraph generalized Gilbert's formulation of these criteria, so the compounded tensor formulation generalizes Kalman's formulation. Thus, if we consider a single layer of the component array of Z, represented by the J'th power (25) of the compounded matrix S, it is not difficult to recognize the symmetrically placed, distinct submatrices $(A^{J-1} \times B)$ and $(C \times A^{J-1})$ making up the general terms of the rank criteria (I.5) and (I.6) for respectively state controllability and observability. Also, as the nucleus of the tensor Z we recognize in (25) the submatrix A^J in terms of which the tensor polynomial Q of (I.12) is defined. In addition, however, we identify a new distinct submatrix $(C \times A^{J-2} \times B)$. It is a peculiar characteristic of the tensor subpolynomial, specified by the latter matrix, that its first non-zero term, defined by the first power (21) of S, is the coefficient matrix D which otherwise does not enter the general expression determining this new matrix. Thus, a total of two null matrices will lead this tensor subpolymial if it so happens that D is a null matrix. As we shall see the identification of the new tensor subpolynomial, gives rise to a formulation of the concept of output controllability in a manner consistent with the derivation of the concepts of state controllability and observability. Inherent in this formulation, to be sure, is the intrinsic property that matrix D must appear as a term of the defining tensor subpolynomial.

Basically, a system is completely output controllable if it is possible, by means of the control variables, to drive in finite time between two specified values, not an arbitrary state variable, but an arbitrary output variable. The conventional matrix formulation of the necessary and sufficient condition that a system is *completely output controllable*, is obtained from the state and output equations (I.1) and (I.2) in a form similar to that of the criteria (I.5) and (I.6). That is, establishing by catenation a new matrix T of dimensions $(\underline{Y}, (\underline{P}-1) \times \underline{U})$ with $\underline{P}$ defined by (27) and (28):

$$T = \{D, C \times B, C \times A \times B, C \times A^2 \times B, \ldots, C \times A^P \times B\} \tag{29}$$

it is required that the rank of this matrix is $\underline{Y}$.

Clearly, to make the expression (29) consistent with the tensor subpolynomial defined by Z, we must extend it by a leading null matrix. But that implies that it actually contains $\underline{P}$ catenated submatrices rather than $(\underline{P}-1)$. Hence, the traditional formulation of complete output controllability confirms the observation made previously that, relative to the establishment of the Cartesian tensor Q, two additional power terms are required in the formulation of the universal tensor Z. This explains why the highest power P of matrix A in (29) must be (X-1), in conformity with the criteria (I.5) and (I.6), in spite of the fact that matrix D is introduced as an additional term. In this connection it is a striking fact that, to be consistent with their derivation from the universal ternsor as their common origin, all three criteria should be established with a null matrix as their initial term. Of course, from a more pragmatic point of view this generalization will be of no avail.

In passing, before we go into a more detailed discussion of the concept of output controllability, it may be of interest briefly to consider the importance of the pattern of zeros in the compounded system matrix S of (21). It is evident, as may be illustrated by considering the aggregated digraph of Fig. 2, that the zero pattern may be explained by the fact that the input or control variables appear as sources whereas the output variables play the role of sinks.

If this picture is disturbed, for example, by introducing a submatrix W into (21) as follows:

$$S = \begin{bmatrix} 0 & C & D \\ 0 & A & B \\ 0 & 0 & W \end{bmatrix} \qquad (30)$$

it is noteworthy, as may be confirmed by a little computation, that this extension will drastically change the formulations of the two criteria of state controllability and output controllability. In fact, only the criterion of observability may be applied in its conventional form, whereas the other two must be restated in far more complicated forms. Based on observations of this nature, therefore, it is not difficult to see why, relative to the pattern of the compounded matrix (21), the combined state and output equation format has gained such a wide acceptance as the preferred way in which to represent the dynamic behaviour of a system.

7.2 On the Track of a New Duality

A reformulation of the criterion of output controllability in terms of Cartesian subtensors of the universal tensor Z, will appear as a straight-forward generalization of the tensor criteria of state controllability and observability provided we can assume that the coefficient matrix D of the output equation is a null matrix. When this is so, by analogy, the natural starting point is a description of the system by the Cartesian tensor Q of (I.12). The combination of the system with its indicator and its control is depicted algebraically by, what we shall call, the *interterminal tensor* T of rank 3:

$$T \leftarrow C + . \times Q + . \times B \qquad (31)$$

With the dimensions of Q given by (I.11), of C given by (I.33), and of B given by (I.15), it is easy to see, as illustrated in Fig. 3, that the dimensions of T (for terminal) are:

$$(\underline{Y}, \underline{P}, \underline{U}) = \rho T \qquad (32)$$

where $\underline{P} = \underline{X}$ by (I.3) and (27).

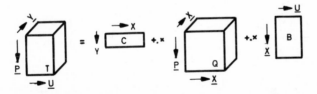

Fig. 3. Component Arrays of the Interterminal Tensor

Thus, if D is a null matrix we may adopt an approach which, by neglecting D, may be explained without taking the universal tensor Z into consideration. Though, it is readily appreciated that the analogy on which this approach is based, will fail if we cannot assume that D is a null matrix. In fact, if D is non-zero it represents the external interconnections between the outputs and the inputs. Hence, to accomodate a non-zero D matrix depicting the invariant physical property of input-output interconnections, it will be necessary to revise our approach in the light of the existence of the universal tensor Z.

By the definition of the universal tensor Z it was established that the D matrix is the initial non-zero layer of the tensor subpolynomial which, by now, may be identified with the generalized form of the interterminal tensor T. Evidently, since the dimensions of matrix D are:

$$(\underline{Y}, \underline{U}) = \rho D \tag{33}$$

the generalized form of tensor T may be established placing D in front of the component array of T specified by (31). In APL, assuming 1-origin, this is done by a catenation of the two arrays:

$$T \leftarrow D, [2] T \tag{34}$$

This operation is illustrated in Fig. 4 from which we derive the dimensions of the resulting tensor T:

$$(\underline{Y}, (\underline{P}+1), \underline{U}) = \rho T \tag{35}$$

where, by (32), $\underline{P} = \underline{X}$. It stands to reason that the catenation (34) and the fact that its outcome is a tensor, can be explained only by reference to the specification of T as a tensor subpolynomial of the universal tensor Z. Therefore, a single rank test will suffice for complete output controllability in-

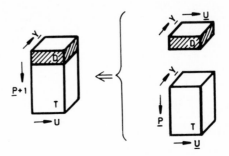

Fig. 4. The Cartesian Tensor Operation $D, [2] T$

dependent of whether the interterminal tensor T is given by (34) or by its simplified form (31).

Basically, the rank test for complete output controllability is performed entirely in agreement with the approach adopted previously for state controllability. Thus, to begin with we establish the 4-dimensional component array of the inner product of T and its generalized transpose $\lozenge T$:

$$T + . \times \lozenge T \tag{36}$$

with the dimensions of this product given by:

$$(\underline{Y}, (\underline{P}+1), (\underline{P}+1), \underline{Y}) = \rho T + . \times \lozenge T \tag{37}$$

if we assume D non-zero and, hence, T specified by (34).

Introducing a tensor contraction on the repeated index $(\underline{P}+1)$ we redefine T as follows:

$$T \leftarrow +/[3] 1 \; 3 \; 3 \; 2 \lozenge T + . \times \lozenge T \tag{38}$$

That is, we obtain a tensor of rank 2 with the dimensions:

$$(\underline{Y},\underline{Y}) = \rho T \tag{39}$$

In other words, the resulting component array is a square matrix with its dimensions specified by the number of output variables. It follows that the rank test for complete output controllability:

$$\rho \boxminus T \tag{40}$$

is simply a matter of determining whether the determinant of T is non-zero.

The crucial observation in connection with this procedure, is that, in order to satisfy the conventional matrix condition (29) for complete output controllability, we are forced to adopt the Cartesian tensor formulation of the rank test for state controllability. In fact, to obtain a square matrix with its dimensions (39), specified by number $\underline{Y}$ of output variables, it is required by necessity that we define the inner product (36) of T with its transpose by the same form as that introduced in (I.18) for the state controllability tensor K. Evidently, *by adopting this form we introduce an asymmetry indicating that, to restore the symmetry, there must exist an alternative rank test* based on the form (I.36) introduced for the observability tensor M. It is immediately seen that these two rank tests are absolutely independent of each other in virtue of the fact that the hitherto neglected rank test is concerned with a square matrix, the dimensions of which are specified by the number $\underline{U}$ of control variables.

To be specific, let us introduce, by analogy with the form of (I.36), the dual of the inner product (36):

$$(\Diamond T)+.\times T \tag{41}$$

with its dimensions specified by:

$$(\underline{U},(\underline{P}+1),(\underline{P}+1),\underline{U}) = \rho(\Diamond T)+.\times T \tag{42}$$

Comparison of this expression with that of (37) indicates that the duality is purely a relationship between input and output variables since, by (35) the interterminal tensor T is only implicitly dependent upon the state variables.

Introducing, corresponding to (38) a tensor contraction on the repeated index (P+1) the interterminal tensor T is redefined:

$$T\leftarrow+/[3]1\ 3\ 3\ 2\Diamond(\Diamond T)+.\times T \tag{43}$$

the result of which is a tensor for rank 2 with the dimensions:

$$(\underline{U},\underline{U}) = \rho T \tag{44}$$

Thus, in contradistinction to (39) the dimensions of the resulting square matrix component are specified by the number of output variables. Determination of whether the determinant of T, given by (43) is non-zero:

$$\rho \boxminus T \tag{45}$$

gives rise to a dual rank test restoring the symmetry of the formulation.

Clearly, by these very formal symmetry considerations we have derived a rank test for a condition dual to that of complete output controllability. The characteristic features of the new rank test are mainly two. First, the form of the rank test (41) was established by analogy with the form of the rank test

(I.36) defining complete observability. Secondly, the dimensions of the resulting square matrix (44) were determined by the number of input or control variables. Since both features are the duals of those defining output controllability the new dual property will be designated *complete input observability*. This name, as will be explained later from a structural point of view, turns out to give an apt description of the intuitive content of the new concept the essence of which, is whether the measurements of the output Y contain sufficient information to identify, in finite time, any input or control variable U. In other words, a system is *completely input observable* if every input of the system affects some of the outputs. Whether or not this concept has more than theoretical interest (e.g. in relation to measurements of unintended inputs such as noise), remains to be seen.

The power of the universal tensor Z, is that it describes the total system in an aggregated format the symmetries of which disclose the invariant properties defining the four concepts of controllability and observability. Thus, considering a single layer (25) of the universal tensor Z in comparison with the compounded system equation (20), it becomes evident how these four concepts relate the variables from one side of (20) to the other. Indeed, since the basic purpose of these concepts is to establish relationships that, given one set of variables, will permit the determination of another set of variables, it is not surprising that all four concepts may be derived from a single tensor expression. Previously, in Table I.1 we listed the dual properties of state controllability and observability. Here, for comparison and ready reference the dual properties of output controllability and input observability have been summarized in Table 2.

OUTPUT CONTROLLABILITY	INPUT OBSERVABILITY
$T \leftarrow C + . \times Q + . \times B$ $(\underline{Y}, \underline{P}, \underline{U}) = \rho T$	
$(\underline{Y}, \underline{U}) = \rho D$ $T \leftarrow D , [\,2\,]T$ $(\underline{Y}, (\underline{P}+1), \underline{U}) = \rho T$	
$(\underline{Y}, (\underline{P}+1), (\underline{P}+1), \underline{Y}) = \rho T + . \times \emptyset T$ $T \leftarrow +/[3]1\ 3\ 3\ 2 \emptyset T + . \times \emptyset T$ $(\underline{Y}, \underline{Y}) = \rho T$	$(\underline{U}, (\underline{P}+1), (\underline{P}+1), \underline{U}) = \rho (\emptyset T) + . \times T$ $T \leftarrow +/[3]1\ 3\ 3\ 2 \emptyset (\emptyset T) + . \times T$ $(\underline{U}, \underline{U}) = \rho T$
$\rho \boxplus T$	$\rho \boxplus T$

Table 2. Output-Input Cartesian Tensor Approach

To illustrate quantitatively the dual concepts of output controllability and input observability let us consider a simple example based on the system properties defined numerically by the Cartesian tensor Q of (I.14). Obviously, the first step is to establish the interterminal tensor T of (31). Let us assume, to emphasize that in general U≠Y, that B is a control *vector* and C is an indicator *matrix*. In this case, the APL implementation will cause a slight complication because, in contradistinction to conventional mathematics, APL distinguishes between a vector and a matrix with a single row or column. To overcome this problem without introducing any unnecessary revisions of the approach summarized in Table 2, we shall find it convenient from an APL point of view to define B as a single-column matrix, say:

$$
\begin{array}{c}
□←B←B3 \\
\begin{array}{c}
0 \\
0 \\
0 \\
2
\end{array} \\
\rho B \\
4\quad 1
\end{array}
\tag{46}
$$

Simultaneously, let the matrix indicator be:

$$
\begin{array}{c}
□←C←C3 \\
\begin{array}{cccc}
\overline{0}.5 & 0 & 0 & 0 \\
0 & 0 & 0.5 & 0
\end{array} \\
\rho C \\
2\quad 4
\end{array}
\tag{47}
$$

Applying (31) the interterminal tensor T is determined:

$$
\begin{array}{c}
T←C+.×Q+.×B \\
\rho T \\
2\quad 4\quad 1
\end{array}
\tag{48a}
$$

which, most simply, may be taken out for inspection along the dimension P̲:

$$
\begin{array}{c}
2\quad 1\quad 3\lozenge T \\
\begin{array}{c}
0 \\
0 \\
\\
0 \\
1 \\
\\
0 \\
0 \\
\\
0 \\
5
\end{array}
\end{array}
\tag{48b}
$$

Let us test this total configuration, assuming D=0, to see whether it satisfies the conditions of complete output controllability and complete input observability.

Starting with the condition for output controllability, the inner product of (36) exhibits, corresponding to (37), the following dimensions:

$$\rho T+.×\mathbb{Q}T$$
$$2 \quad 4 \quad 4 \quad 2 \tag{49}$$

Contraction on the repeated index P=4 of this product yields by (38) the following matrix the dimensions of which are given by (39):

$$□←T←+/[3]1 \ 3 \ 3 \ 2\mathbb{Q}T+.×\mathbb{Q}T$$
$$\begin{array}{ll} 0 & 0 \\ 0 & 26 \end{array} \tag{50}$$
$$\rho T$$
$$2 \quad 2$$

Submitting this matrix to the rank test of (40):

$$\rho \boxplus T$$
$$DOMAIN \ ERROR$$
$$\rho \boxplus T \tag{51}$$
$$\wedge$$

reveals the obvious fact that the system is *not* completely output controllable.

Turning to the condition for input observability we find, based on (48a), that the inner product of (41) is characterized by the following dimensions identified by (42):

$$\rho(\mathbb{Q}T)+.×T$$
$$1 \quad 4 \quad 4 \quad 1 \tag{52}$$

Performing by (43) a contraction on the repeated index P=4 of this product, produces a matrix of a single element the dimensions of which are given by (44):

$$□←T←+/[3]1 \ 3 \ 3 \ 2\mathbb{Q}(\mathbb{Q}T)+.×T$$
$$26 \tag{53}$$
$$\rho T$$
$$1 \quad 1$$

Evidently, applying the rank test (45) to this matrix:

$$\rho \boxplus T$$
$$1 \quad 1 \tag{54}$$

we obtain that the system is completely input observable.

Hence, we have found that with D=0 the total configuration is completely input observable, but not completely output controllable. To remedy the latter situation let us redesign the total configuration. That is, let us revise the interterminal tensor T introducing an interconnection between the control variable and one of the output variables. Mathematically, this interconnection is represented by an additional layer which, placed in front of the component array of T, is specified by a non-zero matrix D with its dimensions given by (33). Assuming D to be a single-column matrix:

$$\square \leftarrow D \leftarrow D1$$

2
0
　　　ρD
2　1

(55)

the revised interterminal tensor T is established from (48a) by (34):

$$T \leftarrow D,[2]T$$
　　ρT
2　5　1

(56a)

where, by (35), we see that the dimension $\underline{P}=4$ has been increased by one to 5 as substantiated by comparison with (48b) of the following print-out:

2　1　3$\lozenge T$
2
0

0
0

0
1

0
0

0
5

(56b)

Testing the revised interterminal tensor T for output controllability we find, corresponding to (37), that the dimensions of the inner product (36) are:

$$\rho T+.\times \lozenge T$$
2　5　5　2

(57)

Submitting this product to the tensor contraction (38) on the repeated index $(\underline{P}+1)=5$, yields a matrix with its dimensions specified by (39):

$$\square \leftarrow T \leftarrow +/[3]1 \; 3 \; 3 \; 2 \lozenge T+.\times \lozenge T$$
4　　　　　　0
0　　　　　26
　ρT
2　2

(58)

It is immediately seen now, as verified by the rank test (40):

$$\rho \boxplus T$$
2　2

(59)

that introducing the non-zero D of (55), has accomplished a revision of the total configuration which made it completely output controllable.

In this connection it should be mentioned that, as implied by the nature of the universal tensor, the introduction of a non-zero matrix D cannot destroy already existing properties of complete output controllability or complete input observability. Accordingly, by (54) the revised total configuration will be completely input observable also.

7.3 Output Controllability and Input Observability

Previously, the dual concepts of output controllability and input observability were considered quantitatively from a rather formal point of view. The point now is to obtain insight into their underlying qualitative characteristics. Thence, our aim is to obtain a formulation of their structural properties of reachability and term rank in a manner comparable to that which was propounded for state controllability and observability. This time, however, we may elide the intermediate step in terms of Boolean tensors and go directly to the formulation based on the computationally far more advantageous reachability matrix. Similarly, only the formulation of the term rank criteria itself will be of interest since we have already dealt with the problem of how to determine term rank by the alternating path method. Thus, beginning with reachability we shall focus on the mathematical formulation in APL postponing the numerical examples to the next section.

By comparison with the approach in section (II.2.2) it is readily visualized that the Boolean design matrix, representing the structural reachability properties of the interterminal tensor T of (31), assuming D to be a null matrix, must be:

$$T \leftarrow C \vee . \wedge R \vee . \wedge B \tag{60}$$

with the dimensions:

$$(\underline{Y}, \underline{U}) = \rho T \tag{61}$$

Now, if the total configuration is expanded by some external interconnections between outputs and inputs, it will be necessary to revise T, specified by (60), to take this extension into account. From the viewpoint of design these external interconnections of the terminals serve the purpose of adjusting the structural properties of the total configuration. To fix ideas, therefore, we shall term these interconnections the *terminal adjuster* in order to emphasize that we consider them a structural unity. Accordingly, *the structural constituents of the total configuration are: the system itself; its control; its indication for output measurements; and its terminal adjuster.*

Basically, the structural properties of the terminal adjuster are depicted by a Boolean matrix D:

$$D \leftarrow D \neq 0 \tag{62}$$

with its dimensions specified by (33).

The revision of the expression (60), made necessary if the total configuration is extended by the terminal adjuster, may be written:

$$T \leftarrow D \vee T \tag{63a}$$

the interpretation of which, invoking (60), is:

$$T \leftarrow D \vee C \vee . \wedge R \vee . \wedge B \tag{63b}$$

Thus, whether or not D is zero (depending upon whether or not a terminal adjuster has been introduced), we end up with a Boolean design matrix T the dimensions of which are specified by (61). It should be emphasized, however, that even if (63b) here is formulated as the Boolean union of two separate terms it depicts in fact a logical summation or reduction by the existential quantifier $\vee/...$ along the dimension (P+1) of a single Boolean tensor polynomial representing the Cartesian interterminal tensor on the left hand side of Fig. 9. In this tensor matrix D appears as a layer in front of the layers defined by the powers of matrix A. It follows that the union of the two terms of (63a) specifies the structural counterpart of the catenation operation in (34).

The interesting fact about the matrix T of (63) and its transpose, is that they define the reachability properties both ways over the distances between input and output terminals of the total systems digraph. The matrix T, therefore, combines in a single array all the reachability properties pertaining to the dual concepts of output controllability and input observability. Truly, it is this fact that separates these concepts from those of state controllability and observability. Also, it is due to this fact that input observability happened to be a hitherto neglected symmetry.

Designating as usual by the term *potential* the structural properties of our quantitative concepts, it is now possible to establish the reachability conditions for potential output controllability and potential input observability by analogy with the corresponding conditions for potential state controllability and potential observability respectively. Reachability, as we have seen, is a structural property completely independent of that of term rank. Hence, the following formulation of the *reachability criteria* for output controllability and input observability must be complemented by the term rank criteria to be specified later.

Considering the concept of *potential output controllability* we find that, in each column, T lists by 1's those outputs that are directly controlled by the input in question. Simultaneously, a *zero row* in this matrix will signify an output that is *not* topologically or structurally related to any of the inputs. Accordingly, in such cases it will *never be possible* to control that output by the given set of inputs. If, alternatively, no zero row appears in the design matrix T the topological structure of the total configuration has the *potential* for complete output controllability provided it also satisfies the corresponding term rank criterion. Whether or not this potential is actually used to attain this property, depends upon the quantitative values assigned and must be decided upon by the rank test of (40).

It follows from this description that if we apply the existential quantifier to the columns of T:

$$T \leftarrow \vee / T \tag{64}$$

we obtain a vector of dimensions $\underline{Y}$:

$$(\underline{Y}) = \rho T \tag{65}$$

Clearly, this vector (64) must be a full unit vector in order that the total configuration is potentially output controllable. Corresponding to the test (I.64) we may investigate this property by the expression:

$$. + / T \tag{66a}$$

the formally correct statement of which, of course, should be in terms of the universal quantifier:

$$\wedge / T \tag{66b}$$

Considering the dual concept of *potential input observability* we find that, in each row, the Boolean design matrix T lists by 1's those inputs that affect the output measurement in question. It follows that a *zero column* in this matrix will signify an input that is *not* topologically or structurally related to any of the output variables. Accordingly, in such cases it will *never be possible* to observe that input by the given set of outputs. If, alternatively, no zero column appears in this matrix the total configuration has the *potential* for complete input observability provided, of course, that also the corresponding term rank criterion is satisfied. Again, whether or not this potential is actually used to attain this property, depends upon the quantitative values assigned and must be decided upon by the rank test (45).

From this description, assuming 1-origin, it follows that if we submit the rows of T to the existential quantifier:

$$T \leftarrow \vee / [1] T \tag{67}$$

the obtained vector of dimensions $\underline{U}$:

$$(\underline{U}) = \rho T \tag{68}$$

must be a full unit vector in order that the total configuration is potentially input observable. To decide upon this matter the vector T of (67) may be submitted to the test:

$$+ / T \tag{69a}$$

or formally more correct, but less informative:

$$\wedge / T \tag{69b}$$

A summary of the dual procedures for determination of the reachability properties of potential output controllability and potential input observability is given in Table 3. The reader may find it of interest to compare the entries of

OUTPUT CONTROLLABILITY	INPUT OBSERVABILITY
$T \leftarrow C \vee . \wedge R \vee . \wedge B$ $(\underline{Y}, \underline{U}) = \rho T$	
$(\underline{Y}, \underline{U}) = \rho D$ $T \leftarrow T \vee D$ $(\underline{Y}, \underline{U}) = \rho T$	
$T \leftarrow \vee / T$ $(\underline{Y}) = \rho T$	$T \leftarrow \vee / [1] T$ $(\underline{U}) = \rho T$
$+ / T$	$+ / T$

Table 3. Output-Input Reachability Matrix Approach

this table with those of Table II.1 defining the reachability properties of potential state controllability and potential observability. Incidentally, to avoid any terminological confusion, it may perhaps be advantageous in general to rename the lastmentioned concept: *potential state observability* (respectively *state observability* if we refer to Table I.1) so as to clearly distinguish it from the new duality of input observability.

With the reachability criteria thus established let us turn to the problem of determining the corresponding *term rank criteria*. Evidently, since potential controllability, as well as potential observability, are derived from identically the same tensor subpolynomial, namely the Boolean interterminal tensor T of (63), only a single term rank need to be determined. Of course, the criteria to be satisfied by this term rank, say W, will in general differ as they depend upon the number of output variables $\underline{Y}$ or the number of input or control variables $\underline{U}$. Thus, assuming the common term rank W to be known, we have that *potential controllability* must satisfy:

$$\underline{Y} \le W \tag{70}$$

whereas *potential input observability* requires:

$$\underline{U} \le W \tag{71}$$

Perhaps, the most striking property of the interterminal tensor T, relative to the state controllability tensor K and the observability tensor M, is the fact that all its layers, defined by powers higher than one of the universal tensor Z, are completely independent of its initial layer defined by matrix D. To be sure, it was this fact that led us to the separation of D from the remainder of T in the quantitative expression (34) and the qualitative expression (63). The implication of this is that we may decompose the term rank determination into two independent parts the term ranks of which are added to produce the common term rank W. One part is the matrix D with its term rank determined by the statement:

$$+/+/\underline{ASSIGN}\ D \tag{72}$$

The other part is the expression (31) for the remainder layers of T. By analogy to the approach adopted in part II this term rank is found by considering the catenation of matrices B and C:

$$+/+/\underline{ASSIGN}\ B,\lozenge C \tag{73}$$

with the dimensions of the catenated matrix given by:

$$(\underline{X},\underline{U}+\underline{Y}) = \rho B,\lozenge C \tag{74}$$

The common term rank W, therefore, is produced adding the results of (72) and (73):

$$W \leftarrow (+/+/\underline{ASSIGN}\ D)++/+/\underline{ASSIGN}\ B,\lozenge C \tag{75}$$

Hence, the term rank criteria for potential output controllability and potential input observability are established combining (75) with respectively (70) and (71). The resulting dual approaches are summarized in Table 4.

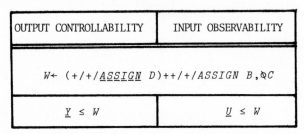

OUTPUT CONTROLLABILITY	INPUT OBSERVABILITY
$W \leftarrow (+/+/\underline{ASSIGN}\ D)++/+/ASSIGN\ B,\lozenge C$	
$\underline{Y} \le W$	$\underline{U} \le W$

Table 4. Output-Input Term Rank Approach

The expression (75) for determination of the common term rank W is interesting for two reasons. First, by the fact that it is derived by a summation, we see that we may indeed consider the terminal adjuster, specified by matrix D, as an independent structural unity. Thus, if D is a null matrix the common term rank W, corresponding to the expression (60), is specified by (73). Secondly, by comparison of (31) and (60) with the argument of (73), it is evident that neither the system tensor Q nor its corresponding reachability matrix R will directly influence the determination of the common term rank W. The role of Q, or perhaps rather R, is to provide the indirect interconnection between input and output, but it is the control and the indication for output measurements, represented respectively by matrices B and C, that determine the term rank. What is provided by the system itself represented by matrix R, is the necessary reachability property. Again, therefore, we see that even if reachability and term rank are independent structural properties, the former must be characterized as more general than the latter.

7.4 A Few Computational Experiments

To familiarize ourselves with the dual concepts of output controllability and input observability these concepts will now be discussed within the framework of experimental computing. That is, we shall consider for these concepts the computational procedures underlying the reachability test of Table 3 and the term rank test of Table 4. By virtue of the fact that these two tests are independent of each other we shall find it convenient to examine the two tests apart. Thus to begin with we shall study the reachability properties of the two concepts for a system satisfying the term rank test. Later, reversing the approach, we shall consider the term rank properties of the two concepts for a system satisfying the reachability test. In both cases, as we shall see, a failure of a system to meet the requirements of one of these tests, calls for a redesign of the system to remedy the situation.

To comply with the illustrative computations used previously let us adopt the system configuration defined numerically in section 7.1. For this configuration the system properties were specified quantitatively by the Cartesian tensor Q of (I.14). The underlying topological structure was expressed by the reachability matrix R of (II.5) or (II.6):

$$
\begin{matrix}
& & R & & \\
1 & 1 & 0 & 0 \\
0 & 1 & 0 & 0 \\
1 & 1 & 1 & 1 \\
1 & 1 & 1 & 1 \\
& & & \rho R \\
4 & 4 & & &
\end{matrix}
$$

Assigning truth-values to the control of (46) and the indicator (47) we find that the corresponding structural properties are:

$$
\begin{matrix}
& \square \leftarrow B \leftarrow B3 \neq 0 & \\
0 & & \\
0 & & (76) \\
0 & & \\
1 & & \\
& \rho B & \\
4 \quad 1 & &
\end{matrix}
$$

$$
\begin{matrix}
& \square \leftarrow C \leftarrow C3 \neq 0 & \\
1 & 0 & 0 & 0 \\
0 & 0 & 1 & 0 & (77) \\
& & \rho C & \\
2 \quad 4 & & &
\end{matrix}
$$

To begin with, assuming that no terminal adjuster has been introduced (i.e., D=0), let us verify the basic assumption that the term rank test is satisfied. Clearly, submitting the argument $B, \diamond C$ to the APL function *ASSIGN* defined in part II, the resulting maximal permutation matrix must be the argument itself:

$$\underline{ASSIGN} \; B, \diamond C$$

$$
\begin{array}{ccc}
0 & 1 & 0 \\
0 & 0 & 0 \\
0 & 0 & 1 \\
1 & 0 & 0
\end{array}
\tag{78}
$$

Hence, by (73) the terminal rank W is:

$$\square \leftarrow W \leftarrow +/+/\underline{ASSIGN} \; B, \diamond C \tag{79}$$
$$3$$

Since, by (77) and (76) the number of output variables $\underline{Y}$ respectively the number of input or control variables $\underline{U}$ are respectively:

$$\square \leftarrow \underline{Y} \leftarrow 1 \uparrow \rho C$$
$$2$$
$$\square \leftarrow \underline{U} \leftarrow {}^-1 \uparrow \rho B \tag{80a\&b}$$
$$1$$

we see by the criteria of Table 4 that indeed the term rank tests for output controllability and input observability are fulfilled:

$$\underline{Y} \leq W$$
$$1 \tag{81a\&b}$$
$$\underline{U} \leq W$$
$$1$$

Turning to the reachability test we find by (60) that the Boolean design matrix T, representing the reachability properties of the total configuration, must be:

$$\square \leftarrow T \leftarrow C \vee . \wedge R \vee . \wedge B$$
$$
\begin{array}{c}
0 \\
1
\end{array}
$$
$$\rho T \tag{82}$$
$$2 \; 1$$

with its dimensions defined by (61).

Evidently, since the first row of this matrix T is zero the reason why this configuration was found not completely output controllable (51), is simply that it does not have the structural potential for being so. Thus, taking the Boolean sum or union of the columns by (64) we find the vector of dimension $\underline{Y}$:

$$\square \leftarrow T \leftarrow \vee /T$$
$$0 \; 1$$
$$\rho T \tag{83}$$
$$2$$

In virtue of the fact that this vector is not a full unit vector, as may be verified by (66a):

$$+/T$$
$$1 \tag{84}$$

this configuration is *not* potentially output controllable.

On the other hand, if we investigate the total configuration for the structural property of being potentially input observable it is immediately seen, since the Boolean design matrix of (82) has only non-zero columns, that it does indeed possess this property. Taking, by (67), the Boolean sum or union of the rows of (82):

$$\square \leftarrow T \leftarrow \vee / [1] T$$

$$\begin{array}{l} 1 \\ \quad \rho T \\ 1 \end{array} \tag{85}$$

this may be verified by the fact that the test (69a):

$$+/T$$

$$1 \tag{86}$$

results in the dimension $U=1$ (see (80b)) indicating a full unit vector albeit degenerated into a single element.

The next step in the numerical example of section 7.1 was to introduce a terminal adjuster defined by the Cartesian matrix D of (55). Correspondingly, turning D into a Boolean matrix we find by (62):

$$\square \leftarrow D \leftarrow D1 \neq 0$$

$$\begin{array}{l} 1 \\ 0 \\ \quad \rho D \\ 2 \ 1 \end{array} \tag{87}$$

Evidently, this does not affect the term rank test (81) since by (75) we simply increase the term rank W of (79) by one:

$$\square \leftarrow W \leftarrow (+/+/\underline{ASSIGN}\ D) + W$$

$$4 \tag{88}$$

On the other hand, it does influence the reachability properties repairing the defect that made the system not potentially output controllable. Thus, by (63a) the Boolean design matrix T, representing the resulting revised configuration, is:

$$\square \leftarrow T \leftarrow D \vee T$$

$$\begin{array}{l} 1 \\ 1 \\ \quad \rho T \\ 2 \ 1 \end{array} \tag{89}$$

Obviously, since both of the rows of this matrix are non-zero the revised configuration is potentially output controllable. As before this may be verified by forming the vector T of (64):

$$\square \leftarrow T \leftarrow \vee / T$$

$$\begin{array}{l} 1 \ 1 \\ \quad \rho T \\ 2 \end{array} \tag{90}$$

and executing the test (66a):

$$+/T$$

$$2 \tag{91}$$

indicating that T is a full unit vector.

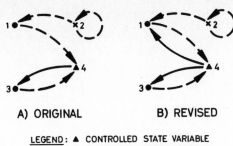

A) ORIGINAL   B) REVISED

LEGEND: ▲ CONTROLLED STATE VARIABLE
         ● OBSERVED STATE VARIABLE
         ✗ OTHER STATE VARIABLE

Fig. 5. Output Controllability and Input Observability

A digraph interpretation of these results is given i Fig. 5 that may be explained as follows. In Fig. 5A we have depicted the system digraph assuming D=0. Similarly, the digraph of Fig. 5B represents the combined digraph of the system and the terminal adjuster introduced by the non-zero D defined by (87). In either case, consulting (76), we find that the directly controlled state variable is $X[4]$. Correspondingly, we derive from (77) that the two directly observed or measured state variables are $X[1]$ and $X[3]$. Accordingly, in the two digraphs of Fig. 5 we mark vertex 4 as a controlled variable and vertices 1 and 3 as observed variables.

Now, considering the total configuration, assuming D=0, we see from the Boolean design matrix T of (82):

$$\square \leftarrow T \leftarrow C \vee . \wedge R \vee . \wedge B$$

$$
\begin{array}{ll}
0 & \\
1 & \\
 & \rho T \\
2 \ \ 1 &
\end{array}
$$

that the observed vertex 1 (the first row index) is *not* a successor of the controlled vertex 4 (the column index), whereas the observed vertex 3 (the second row index) is indeed a successor. Indicating by heavy lines edges that are defined by 1's in this Boolean design matrix the digraph of Fig. 5A results. It is easy to see then that the Boolean design matrix simply expresses the digraph reachability properties defining the concepts of potential output controllability and potential input observability. Truly, in either case we are concerned with the reachability properties *from* the controlled vertices *to* the observed vertices. Thus, in the case of potential output controllability we demand that *all* observed vertices (here 1 and 3) are reachable from *one or more*, but not necessarily all, of the controlled vertices (here 4). Clearly, this condition is not satisfied by the digraph of Fig. 5A. Alternatively, in the case of potential input observability the requirement is that from *all* controlled vertices (here only 4) *one or more*, but not necessarily all, observed variables (here 1 and 3) should be reached. Obviously, this condition is satisfied by the digraph of Fig. 5A. The abstract "fixed point of reflection" of this symmetry, is the situation where *all* observed vertices are reachable from *all* controlled vertices, because in this situation the dual concepts come together. To illustrate, this will occur in Fig. 5A if vertex 3 is the only observed vertex and vertex 4 the only controlled vertex.

An interesting feature of the Boolean design matrix T of (82) is the fact that it tells us, by its zero rows or zero columns, in which corresponding rows or columns of the Cartesian tensor D that, at least, non-zero entries must be introduced. Evidently, our choice of the Boolean tensor D in (87) was guided by this rule. The digraph interpretation of the result of introducing a non-zero D, is exemplified in Fig. 5B. All this really amounts to is the introduction into the digraph of a set of additional edges defined by D, conceived as a successor matrix. Thus, by D of (87) we introduce the observed vertex 1 as an immediate successor of the controlled vertex 4. The result of this operation is found directly by inspection of the digraph of Fig. 5B. Namely, that *all* the observed vertices (here 1 and 3) are reachable from *some* of the controlled vertices (here the only controlled vertex 4). In other words, the revised configuration is potentially output controllable. It is obvious, in this connection, that if from *all* controlled vertices *one or more* observed variables are reachable, then expansion of the digraph by some additional edges will not destroy this property of potential input observability. This is the topological argument underlying the statement in section 7.2 that the introduction of a non-zero D cannot nullify existing properties of complete output controllability and complete input observability.

Previously, in section 4.1, after having discussed the digraph interpretation of the reachability properties of the concept of potential state controllability we introduced the converse digraph to explain the concept of (state) observability. Above we adopted the more convenient approach in practice of relating our discussion of the dual reachability properties of potential output controllability and potential input observability to only the original or primal successor digraph. Thus, most people usually refer to the original or primal digraph by terms such as *are reachable from* and to the converse digraph by expressions like *from....are reached*. However, this somewhat careless attitude that comes with familiarity with a subject, may tend to confuse the issue in the mind of anyone less acquainted with the topic. Therefore, to make it absolutely clear that the concept of potential input observability is the directional dual of the concept of output controllability, let us briefly demonstrate how the reachability condition of potential input observability is derived from the converse digraph of Fig. 5A. This is shown in Fig. 6 and may be explained as follows.

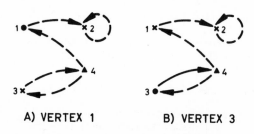

A) VERTEX 1   B) VERTEX 3

LEGEND: ▲ CONTROLLED STATE VARIABLE
        ● OBSERVED STATE VARIABLE
        ✕ OTHER STATE VARIABLE

Fig. 6. Input Observability by the Converse Digraph

Basically, the question of input observability is whether or not, on the converse digraph, all of the controlled vertices, here vertex 4, are reachable from at least one vertex in the set of observed vertices, here vertices 1 and 3. Consulting Fig. 6A which depicts the reachability properties from observed vertex 1, it is immediately seen that controlled vertex 4 is not reachable. On the other hand, inspection of Fig. 6B illustrating the reachability properties from observed vertex 3, reveals that controlled vertex 4 is in fact reachable. Hence, in complete agreement with the earlier statement the configuration satisfies the requirement of input observability.

Now, turning back to Fig. 5B, the basic characteristic of introducing a terminal adjuster, represented by a non-zero D, is the fact that in the digraph it can define edges only from controlled vertices to observed vertices. However, all we really want to do, is to establish the reachability from some controlled vertices to some observed vertices, maintaining the validity of the term rank test. From inspection of the digraph in Fig. 5A it is easy to visualize many other structural possibilities of accomplishing this goal besides that of adding a terminal adjuster to the given configuration.

Assume, for example, that, instead of introducing a non-zero D, we interconnect the two output vertices 1 and 3 so that 1 is the successor of 3. This is illustrated in the digraph of Fig. 7A. Visual inspection of this digraph immediately reveals that this configuration is potentially output controllable as well as potentially input observable. In another situation we may go further than just redesigning the indicator. Thus, according to the dictum that a system should be so designed that it is structurally amenable to control as well as observation, we may choose to introduce directed edges incident on any of the state variable vertices. To illustrate, in Fig. 7B the "internal" vertex 2 is made the immediate successor of the observed vertex 3. Inspection of this digraph readily reveals that it possesses all of the required reachability properties in virtue of forming a socalled *strong component* (i.e., any vertex is reachable from any other vertex).

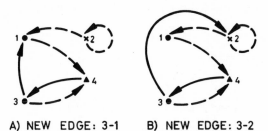

A) NEW EDGE: 3-1   B) NEW EDGE: 3-2

LEGEND: ▲ CONTROLLED STATE VARIABLE
        ● OBSERVED STATE VARIABLE
        × OTHER STATE VARIABLE

Fig. 7. Possible Redesigns

It should be noted, by the way, that the application of digraphs to provide a redesign that will attain the structural reachability properties of any of the four concepts of potential controllability and observability, is but one use of this technique. In fact, another may be to do away with undesirable reachability properties among the state variables. For instance, if in a digraph representation some control vertex depicts a noise input we may use the concept of input observability to pick out an observed vertex which, undisturbed by that noise, may serve, say, as a reference measurement.

But let us reverse our approach by considering the term rank properties of output controllability and input observability for a system satisfying the reachability test. To this end, maintaining as our numerical illustration the system discussed above, let us assume that its indicator of (77) is redefined as follows:

$$
\begin{array}{c}
\square \leftarrow C \leftarrow C4 \neq 0 \\
\begin{array}{cccc} 0 & 0 & 0 & 1 \end{array} \\
\begin{array}{cccc} 0 & 0 & 0 & 1 \end{array} \\
\rho C \\
\begin{array}{cc} 2 & 4 \end{array}
\end{array} \tag{92}
$$

Hence, with the control specified by (76) and assuming the terminal adjuster D=0 the Boolean design matrix T substituting (82) will express the reachability properties of the total configuration as follows:

$$
\begin{array}{c}
\square \leftarrow T \leftarrow C v . \wedge R v . \wedge B \\
1 \\
1 \\
\rho T \\
\begin{array}{cc} 2 & 1 \end{array}
\end{array} \tag{93}
$$

Evidently, since instead of (83) and (84) we now find:

$$
\begin{array}{c}
\square \leftarrow T \leftarrow v / T \\
\begin{array}{cc} 1 & 1 \end{array} \\
\rho T \\
2 \\
\quad +/T \\
2
\end{array} \tag{94a\&b}
$$

the revised configuration satisfies the reachability test for potential output controllability.

Similarly, the reachability test for potential input observability will be complied with since, corresponding to (85) and (86), we have:

$$
\begin{array}{c}
\square \leftarrow T \leftarrow v / [1] T \\
1 \\
\rho T \\
1 \\
\quad +/T \\
1
\end{array} \tag{95a\&b}
$$

With the reachability test thus fulfilled the question of whether or not the revised configuration is potentially output controllable and potentially input observable, is determined solely by the term rank test. With the terminal adjuster D equal to zero the critical argument is:

$$
\begin{array}{c}
B . \& C \\
\begin{array}{ccc} 0 & 0 & 0 \end{array} \\
\begin{array}{ccc} 0 & 0 & 0 \end{array} \\
\begin{array}{ccc} 0 & 0 & 0 \end{array} \\
\begin{array}{ccc} 1 & 1 & 1 \end{array}
\end{array} \tag{96}
$$

Obviously, the maximal permutation matrix contained in (96) is:

$$
\begin{array}{c}
\underline{ASSIGN} \; B . \& C \\
\begin{array}{ccc} 0 & 0 & 0 \end{array} \\
\begin{array}{ccc} 0 & 0 & 0 \end{array} \\
\begin{array}{ccc} 0 & 0 & 0 \end{array} \\
\begin{array}{ccc} 1 & 0 & 0 \end{array}
\end{array} \tag{97}
$$

It follows, therefore, invoking (73) that the term rank W is:

$$\Box \leftarrow W \leftarrow + / + /\underline{\mathit{ASSIGN}} \ B, \!\$C \qquad\qquad (98)$$
$$_1$$

Comparison of (92) and (77) reveals that (80) still holds. Hence, by the term rank tests of Table 4 we find, in contradistinction to (81a), that the system is *not* potentially output controllable since:

$$\underline{Y} \le W \qquad\qquad (99a)$$
$$_0$$

but that it is potentially input observable since, corresponding to (81b), we have:

$$\underline{U} \le W \qquad\qquad (99b)$$
$$_1$$

It is thought-provoking that the deficiency identified by (99a), may be removed by introducing the terminal adjuster D of (87) since, as we saw in (88), D increases the term rank by one:

$$+ / + /\underline{\mathit{ASSIGN}} \ D \qquad\qquad (100)$$
$$_1$$

That is, invoking (75) the term rank W of (98) is revised as follows:

$$\Box \leftarrow W \leftarrow (+ / + /\underline{\mathit{ASSIGN}} \ D) + W \qquad\qquad (101)$$
$$_2$$

Therefore, instead of the term rank test of (99a) we now find:

$$\underline{Y} \le W \qquad\qquad (102)$$
$$_1$$

which shows that the system configuration, augmented by D of (87), is potentially output controllable.

Of course, emulating the approach adopted previously in connection with the failure of the reachability test, we may, instead of redesigning the system, augment it by the terminal adjuster D. Thus, considering solely the term rank test we may conceive, by virtue of (81), the indicator C of (77) as a possible redesign of (92). Evidently in more complicated situations the discovery of a possible redesign may derive from a digraph representation in terms of alternating paths of the system term rank properties. This representation, to be sure, should be taken together with a digraph representation of the system reachability properties in order that the proposed redesign may meet the combined criteria for term rank and reachability.

The usefulness and versatility to be obtained from introducing digraphs into modern control theory stem from the fact that they give us a geometrically simple picture of the structural or qualitative properties underlying any control problem. Conventionally, the quantitative concepts of controllability and observability can guarantee only that a proposed control or indication will serve their purpose, but no clue is given as to how to do it. The digraph interpretation turns these concepts into design tools by providing a geometrical illustration of the structural possibilities inherent in the problem. It is the combination of these possibilities with a physical knowledge of the problem at hand that may provide the heuristic guidelines through the design process. Of course, in this connection we should be aware that controllability and observability, albeit important concepts in modern control theory, only govern the existence of possible solutions. To illustrate, the condition of controllability will decide whether or not a new desirable state is obtainable, but it is the quite different concept of stability that determines the success of actually maintaining that state.

CONCLUSION

Contrasting extreme viewpoints like those of system decomposition and system aggregation, serves the purpose of separating and identifying the fundamental concepts. In this respect, clearly, it was advantageous to tie up the digraph approach with decomposition and the tensor approach with aggregation. Yet, as we have seen in the detailed discussion of this part it is impossible to maintain this basic pattern rigidly. For example, to clarify the role of the universal tensor and derive the concepts of output controllability and input observability from its formulation we found it convenient to expedite the presentation by a suitable digraph representation. Thus, as we have seen also in parts I and II, the tensorial approach and the graph-theoretical approach may be employed alternatively as complementary mathematical tools of representation. Which viewpoint to adopt depends upon the given situation. However, to make a rational choice we must be familiar with both viewpoints.

The decomposition of a system by partitioning of its digraph representation according to quasi-levels, illustrates a general technique the usefulness of which extends far beyond the realm of control theory. In contradistinction the formulation of the universal tensor depends decisively on the fact that the system description can be cast in the format of the state and output equations. Though, when this is so it gives the most compact and simple, mathematical presentation of the inherent symmetries of the system properties of controllability and observability. Of course, whether or not we desire to endow a system design with all of these properties is an entirely different manner. In particular, the usefulness in practice of the new duality of input observability is at present an open question.

Chapter 8. SOME GENERAL REMARKS

The use of digraphs to represent reachability properties of state variables depicted as vertices, is a wellknown technique in the engineering and allied sciences. Some typical applications are, in computer science, finite-state machi-
nes or automata and, in operations research, finite Markov chains. However, a
very important graph-theoretical characteristic of the deterministic automata
and their probabilistic counterparts, is the fact that they completely lack in
their physical or mathematical structure those directional duality properties on
which any discussion of the converse concepts of controllability and observability must be based. In contradistinction, if we consider engineering-economic
production systems we find here this directional duality in the digraph description. The approach in this book originated in the observation that automatic
control systems and engineering-economic production systems exhibit a fundamental analogy which relates input variables to production requirements, output variables to prices, and the state equation coefficient matrix to a bill-of-material
specification.

In principle, if we confine ourselves to the structural properties of some purely
physical configuration, we have been concerned only with two basic concepts of
modern control theory: potential controllability and potential observability.
These concepts, essentially, derive from the topological properties of reachability among the state variables of the given system. A particularly characteristic feature in this respect, is the fact that any discussion of these concepts
must be based on some subdivision into disjoint classes of the total set of state
variables. The kinds of classes of state variables that come into play, vary
from case to case, but in principle we are concerned with only three defining
properties:

- *Controlled state variables*, i.e. directly influenced by some inputs.
- *Observed state variables*,   i.e. directly measured by some outputs.
- *Inaccessible state variables*,i.e. not directly related to inputs or outputs.

Basically, the temporal aspects of the fundamental state and output equations,
(1) and (2), define a causal relationship from the controlled state variables
through the inaccessible state variables and over to the observed state variables. It is the investigation of this causal relationship relative to the
structural properties of the system that we focus upon by the concepts of potential controllability and observability. Considering the digraph representation of the system structure, defined topologically by the immediate successor relationship among its vertices (depicting the state variables), it is
seen that the structural content of this causal relationship derives from the
reachability properties of the digraph.

In particular, following the directed paths of this digraph from any vertex,
designating a controlled state variable, we will discover all the state variables
that are successors of this controlled state variable. That is, we will find
all the state variables which, in virtue of the system structure, may be causally related to the controlled state variable in question. This is the essence
of the concept of *potential controllability*.

Alternatively, if we go in the opposite direction of the directed paths of this
digraph from any vertex, designating an observed state variable, we will discover all the state variables that are predecessors of this observed state variable. That is, we will find all the state variables upon which, in virtue
of the system structure, the observed state variable in question may be causally dependent. This is the essence of the concept of *potential observability*.

It is easy to see that potential controllability and potential observability are directional dual or converse concepts. That is, they exhibit directional dual or converse reachability properties. Therefore, applying one of the concepts to the converse digraph will turn it into the other concept. In other words, only one abstract procedure is needed if we apply it to a digraph and its converse. This was exactly what we did when we investigated the total configuration for potential observability of the converse digraph.

Now, the difference between, on the one hand, the converse concepts of potential state controllability and potential (state) observability and, on the other hand, the converse concepts of potential output controllability and potential input observability, is simply a matter of defining the two classes of state variables of interest. That is, in complete agreement with the previous exposition of the basic converse concepts of potential controllability and potential observability, each of the abovementioned four more specialized concepts is defined in terms of only two classes of state variables.

Potential state controllability is based on a partition of the state variables into the class of controlled state variables and the complement class of all the remaining other state variables. Thus, the latter class may be defined as the union of the class of the observed state variables and the class of the inaccessible state variables. Considering the *original digraph* the condition to be satisfied, is that *all* members of the complement class should be reachable from *at least one* member of the class of controlled variables.

Potential (state) observability is based on a partition of the state variables into the class of observed state variables and the complement class of all the remaining other state variables. Thus, the latter class may be defined as the union of the class of controlled state variables and the class of inaccessible state variables. Considering the *converse digraph* the condition to be satisfied, is that *all* members of the complement class should be reachable from *at least one* member of the class of observed variables.

The characteristic feature of this pair of converse concepts, therefore, is that they are based explicitly on the total set of state variables that are divided into two disjoint classes. In contradistinction the pair of converse concepts of potential output controllability and potential input observability, is based explicitly on *only a subset* of the total set of state variables. That is, this pair of converse concepts is defined in terms of the two classes of respectively controlled and observed state variables only. It is this difference that forces us to distinguish between the two pairs of converse concepts.

Potential output controllability, considering the *original digraph,* is based on the condition that *all* members of the class of observed state variables should be reachable from *at least one* member of the class of controlled state variables.

Potential input observability, considering the *converse digraph,* is based on the condition that *all* members of the class of controlled state variables should be reachable from *at least one* member of the class of observed state variables.

Thus, in the case of the lastmentioned pair of converse output-input concepts we are concerned only with the externally directly accessible state variables. That is, the sole contribution of the inaccessible state variables, is implicitly to provide the reachability properties between the two classes of respectively controlled and observed state variables.

In the context of systems theory we may expect that only the invariant tensorial properties, represented structurally by the reachability of the systems digraph, will be of general interest. This is the reason why in the preceding comparison the concepts of potential controllability and observability, were classified according to their reachability properties. However, confining ourselves to the systems of modern control theory it will be necessary also to consider the term rank properties. By virtue of the fact that the term rank condition is the structural counterpart of Gilbert's assumption, it is readily appreciated that it cannot influence the abovementioned classification. To be sure, term rank is but a non-tensorial property reflecting the structural possibility of the existence of an appropriate matrix inverse. It follows that in the definition of potential controllability and observability term rank will play a secondary role in comparison with reachability.

Two basic requirements, to be met by any scientific explanation, are general usefulness and explicit statement. Essentially, the attitude taken here, is that the first requirement is concerned with the search for order within the given facts while the second is directed towards expressing that order operationally for prediction. Therefore, in writing this book we have endeavoured simultaneously to satisfy two objectives.

The main objective has been to provide a simple geometrical description relating the fundamental concepts of controllability and observability to the underlying structural properties of a control system. In principle this description being based on the reachability properties and maximal alternating paths of a digraph representation of the state variables, is purely topological. That is, we have elided in this description any trace that characterizes the system properties quantitatively by substituting the wellknown flowgraph technique by its digraph counterpart. But it is solely the quantitative attributes of these properties that permit us to distinguish between linear and nonlinear systems. Hence, it is by focusing on the qualitative or topological characteristics of the system structure that we can hope to identify those invariant relationships that may be generalized from the linear to the nonlinear domain. The clarification of the more exact nature of the abstract symmetry between the converse concepts of controllability and observability, including the identification of the neglected duality of input observability, may serve as a first step in this direction.

A secondary objective has been to demonstrate how the APL notation may serve to provide an explicit and unique description, making operational the algebraic formulation, say, from Cartesian tensors over Boolean tensors to the reachability and the associated Boolean design matrices. Of particular interest in this connection is the facility of using the APL terminal as a computational laboratory to carry out numerical experiments very much as we perform physical experiments in a conventional laboratory. Creation of order out of fact does not begin with a set of axioms followed by the development of theorems and lemmas. Rather, we start out by proposing certain hypotheses the inferences of which may be tested experimentally. It is in this light, to try out computationally predictions from plausible assumptions, that we have found the APL terminal a useful engineering tool.

R E F E R E N C E S

Bellman, R.: *Introduction to Matrix Analysis*. McGraw-Hill Inc., N.Y., 1960

Christofides, N.: *Graph Theory - An Algorithmic Approach*. Academic Press, N.Y.,1975

Churchman, C.W., Ackoff, R.L. & Arnoff, E.L.: *Introduction to Operations Research*. John Wiley & Sons, Inc., N.Y., 1957.

Davison, E.J.: "Connectability and Structural Controllability of Composite Systems." *AUTOMATICA*, Vol. 13, 1977, 109-123.

van Dixhoorn, J.J. & Evans, F.J. (eds.): *Physical Structure in Systems Theory - Network Approaches to Engineering Economics*. Academic Press, London, 1974.

Dzubak, B.J. & Warburton, C.R.: "The Organization of Structured Files." *Comm. ACM*, Vol. 8, 1965, 446-452.

Ford, L.R. & Fulkerson, D.R.: *Flows in Networks*. Princeton University Press, Princeton, N.J., 1962.

Franksen, O.I.: "A Point Set Algebra of File Operations." Electric Power Engineering Department, The Technical University of Denmark, March, 1968.

Franksen, O.I.: "On the Future in Automatic Control Education." *AUTOMATICA*, Vol. 8, 1972, 517-524.

Franksen, O.I.: "On Measurements and Their Group-Theoretical Foundation." *Interdisciplinary Conference on Problem Analysis in Science and Engineering*. University of Waterloo, Canada, 1975. Reprinted in Branin, F.H. & Huseyin, K. (eds.): *Problem Analysis in Science and Engineering*. Academic Press, N.Y., 1977.

Franksen, O.I.: "Mr. Babbage, The Problem of Notation, and APL." Lecture Notes. Electric Power Engineering Department, The Technical University of Denmark,1976.

Franksen, O.I., Falster, P. & Evans, F.J.: "Potential Controllability and Observability." Electric Power Engineering Department, The Technical University of Denmark. Publ. No. 7608, June, 1976.

Franksen, O.I.: "Experimental Computing - A Case Study Using APL." April, 1977 (To appear).

Franksen, O.I.: "Group Representations of Finite Polyvalent Logic - A Case Study Using APL Notation." *IFAC VII World Congress*, 1978, Helsinki, Vol.2,875-887.

Gilbert, E.G.: "Controllability and Observability in Multivariable Control Systems." *J. S.I.A.M. Control*, Ser. A. Vol. 1, No. 2, 1963, 128-151.

Glover, K. & Silverman, L.: "Characterization of Structural Controllability." *IEEE Trans. Automatic Control*, Vol. AC-21, No. 4, 1976, 534-537.

Harary, F.: "A Graph Theoretic Method for the Complete Reduction of a Matrix with a View Toward Finding its Eigenvalues." *J. Math. & Phys.*, Vol. 38, 1959, 104-111.

Harary, F.: "A Graph Theoretic Approach to Matrix Inversion by Partitioning." *Numer. Math., Vol. 4, 1962, 128-135.*

Harary, F., Norman, R.Z., & Cartwright, D.: *Structural Models - An Introduction to the Theory of Directed Graphs*. John Wiley & Sons, Inc., N.Y., 1965.

Iri, M., Tsunekawa, J., & Yajima, K.: "The Graphical Techniques used for a Chemical Process Simulator "Juse Gifs"." *Information Processing 71.* North-Holland Publishing Co., 1972, 1150-1155.

Iverson, K.E.: *A Programming Language.* John Wiley & Sons, Inc., N.Y. 1962.

Kalman, R.E.: "Mathematical Description of Linear Dynamical Systems". *J. S.I.A.M. Control,* Ser.A, Vol. 1, No. 2, 1963, 152-192.

König, D.: *Theorie der Endlichen und Unendlichen Graphen.* Chelsea Publishing Company, New York, N.Y., 1950 (Reprint of the original 1936 edition).

Lancaster, P.: *Theory of Matrices.* Academic Press, N.Y., 1969.

Lanczos, C.: *The Variational Principles of Mechanics.* University of Toronto Press, Toronto, 1949.

Lin, C.T.: "Structural Controllability". *IEEE Trans. Automatic Control,* Vol. AC-19, No. 3, 1974, 201-208.

Luce, R.D.: "A Note on Boolean Matrix Theory". *Proc. Am. Math. Soc.,* Vol. 3, 1952, 382-388.

Ore, O.: *Theory of Graphs.* Americal Mathematical Society, Colloquium Publications, Vol. XXXVIII. Providence, R.I., 1962.

Pakin, S.: *APL\360 – Reference Manual.* Science Research Associates, Inc., Chicago, 1968.

Petersen, J.: *De algebraiske Ligningers Theori. (The Theory of Algebraic Equations).* (In Danish). A.F. Høst & Søns Forlag, Copenhagen, 1877.

Petersen, J.: "Die Theorie der Regulären Graphs". *Acta Mathematica,* Vol. 15, 1891, 193-200.

Ryser, H.J.: *Combinatorial Mathematics.* Carus Mathematical Monographs, No. 14. Mathematical Association of America, Inc., 1963.

Shields, R.W. & Pearson, J.B.: "Structural Controllability of Multiinput Linear Systems". *IEEE Trans. Automatic Control,* Vol. AC-21, No. 2, 1976, 203-212.

Siljak, D.D.: "On Reachability of Dynamic Systems". *Int. J. Syst. Sci.,* Vol. 8, No. 3, 1977a, 321-338.

Siljak, D.D.: "On Pure Structures of Dynamic Systems". *Nonlinear Analysis, Theory, Methods & Application,* Vol. 1, No. 4, 1977b, 397-413.

Stevens S.S.: "On the Theory of Scales and Measurements." *Science,* Vol. 3, No. 2684, June 1946, 677-680.

Steward, D.V.: "On an Approach to Techniques for the Analysis of the Structure of Large Systems of Equations". *S.I.A.M. Review,* Vol. 4, No. 4, 1962, 321-342.

Timothy, L.K. & Bona, B.E.: *State Space Analysis: An Introduction.* McGraw-Hill, Inc., N.Y. 1968.

Whittaker, E.T.: *A Treatise on the Analytical Dynamics of Particles and Rigid Bodies,* Cambridge University Press. 4th Edition, 1960.